DMV Seminar
Band 18

Springer Basel AG

Klaus W. Roggenkamp
Martin J. Taylor

Group Rings and Class Groups

Springer Basel AG

Autoren:

K. W. Roggenkamp
Universität Stuttgart
Mathematisches Institut B
Pfaffenwaldring 57
D–7000 Stuttgart 80
Germany

M. J. Taylor
Dept. of Mathematics
U.M.I.S.T.
P.O. Box 88
Manchester M60 1QD
England

Deutsche Bibliothek Cataloging-in-Publication Data

Group rings and class groups / Klaus W. Roggenkamp ; Martin J. Taylor. – Basel ; Boston ;
Berlin : Birkhäuser, 1992
  (DMV-Seminar ; Bd. 18)
  ISBN 978-3-7643-2734-7   ISBN 978-3-0348-8611-6 (eBook)
  DOI 10.1007/978-3-0348-8611-6
NE: Roggenkamp, Klaus W.; Taylor, Martin J.; Deutsche
  Mathematiker-Vereinigung: DMV-Seminar

© 1992 Springer Basel AG
Originally published by Birkhäuser Verlag Basel in 1992

ISBN 978-3-7643-2734-7

Preface

These notes form an extended version of talks given at the DMV–seminar in Günzburg, September 1990. The seminar consisted of two parts :

1) "The isomorphism problem for integral group rings", with the main talks given by K. W. Roggenkamp and shorter contributions by W.Kimmerle, J.Ritter and A. Zimmermann (Part 1).

2) "Galois–Module structure", with the main talks given by M.Taylor and shorter contributions by N.Byott (Part 2).

We greatly appreciate the opportunity, given us by the DMV to hold this seminar.

DMV-Seminar Part 1
Group Rings:
Units and the Isomorphism Problem

K. W. Roggenkamp

with contributions by
W. Kimmerle and A. Zimmermann

Table of Contents

Introduction

The results and arguments in the sections on the isomorphism problem are due to Leonard Scott in collaboration with K. W. Roggenkamp in the last ten years unless otherwise stated [1] .

We have tried to write at least the first sections in great detail, putting together results, and linking different topics which are mostly available only in articles or which are folklore. [2]

In the first two sections we briefly recall some basic properties about modules, blocks, cohomology rings and profinite groups, which will be used in the sequel. We also prove Coleman's lemma, which shows that the normalizer of a p–subgroup modulo its centralizer is determined by the integral group ring.

Sections III, IV and V deal with general properties of finite subgroups in the group of normalized units of integral group rings. We have tried to prove these results in the furthest possible generality, in particular with respect to the choice of the base ring. Here it is shown that the coefficient of the identity of a non trivial unit of finite order in the augmented units is zero, which implies that the elements of a finite subgroup in the normalized units are linearly independent. This result is then used to prove the class sum correspondence between group bases. As a consequence we derive the rigidity of central units of finite order. In section V, which is written by Wolfgang Kimmerle it is shown, that the class sum correspondence is compatible with the pth power map. Moreover, properties of the groups are derived if two groups form a Brauer pair, namely that the character table determines the length of the conjugacy classes and the lattice of normal subgroups. A sketch is given of the theorem that the character table determines the chief series of the group. Finally a short proof of Whitcomb's theorem for metabelian groups is presented and briefly the questions of when the group G has a normal complement in the group of augmented units is discussed.

In section VI the main result on group rings of p–groups is proved, namely that p–adically, finite subgroups of the normalized units are conjugate to a subgroup of the underlying finite group. Here too, the coefficient ring is very general.

In section VII the connection between automorphisms and invertible bi-modules is developed in great generality, namely not only for automorphisms but for injections of finite subgroups into the group of units and is then applied to construct various examples: The involutions in the group ring in the symmetric group of order 6 and the dihedral group of order 8 are calculated.

In section VIII we construct two non–isomorphic metabelian groups of order remarkably smaller than the examples of Dade, which have isomorphic

1   We greatly appreciate the support of the NSF and DFG.
2   I would like to thank Alexander Zimmermann very much for typing most of the manuscript.

group rings at the localization of the integers of every prime. We also give an example, where this holds for certain extension rings of the integers but not for the integers, thus showing that a Noether–Deuring type theorem does not hold for integral group rings.

Section IX is devoted to the Zassenhaus conjecture: that integral group bases are conjugate in the rational group algebra. First we construct a semi local counterexample which is a metabelian even supersolvable group of order 6720. And finally we construct a cocycle in a Čech–type cohomology group which is the obstruction for the weak form of the Zassenhaus conjecture to hold for solvable groups.

Section X is written by Wolfgang Kimmerle and deals with variations on the Zassenhaus conjecture. It is shown that for a direct product of symmetric groups the Zassenhaus conjecture holds, and that for a direct product of alternating groups a variation holds, namely that an isomorphism between the groups can be chosen, which fixes the "class sums" of all cyclic subgroups. Finally in this chapter the connection between the Brauer tree and automorphisms of group rings are derived.

Section XI is written by Alexander Zimmermann and contains parts of the results of his "Diplomarbeit" and of his forthcoming theses, in particular the isomorphism problem for nilpotent on abelian groups is proved as well as a generalization of this result (the main result is stated without proof).

Section XII is again written by Wolfgang Kimmerle and deals with classes of groups which satisfy a certain variation of the Zassenhaus conjecture. In particular he proves that for nilpotent by abelian groups a variation of the Zassenhaus conjecture for p-power elements holds. This yields a positive answer for the the isomorphism problem for groups with nilpotent commutator subgroup, in particular for supersoluble groups. The proof is closely related to the Čech - type cocycles of section IX. The last paragraph of section XII deals with Sylow and Hall subgroups. The main steps for the proof that the character table of a finite group determines, whether or not it has abelian π–Hall subgroups are given. This answers a question of Richard Brauer. The integral group ring however determines even hamiltonian Hall subgroups and nilpotent Hall subgroups for solvable groups.

In section XIII a detailed analysis of Clifford theory even over the p-adic integers (not necessarily a splitting) is developed.

The final section XIV gives various examples of subgroups of the groups of units.

Stuttgart, December 1991

I   Some general facts

§1   Ring reduction to PID

[R-S; 87 1, Appendix]

In order to present the results in their greatest generality, we shall sometimes be using the following reduction:

1.1. Proposition. *Let R be a noetherian integral domain and $\mathcal{P} \subset R$ be an ideal. Then R is contained in a principal ideal domain in which no element of $\mathcal{P}$ is invertible.*

PROOF: For each $x \in \mathcal{P}$ choose a prime ideal $\wp(x)$ containing x and minimal with that property. By the Krull Hauptidealsatz [A-M; 69, 11.7], $\wp(x)$ has height 1 . Let $R_{\mathcal{P}}$ be the localization of R at the multiplicative set $S \setminus (\cup_{x \in \mathcal{P}} \wp(x))$. By [A-M; 69, 3.11] the only non zero prime ideals in $R_{\mathcal{P}}$ are the extensions of the ideals $\wp(x)$. Moreover, R is noetherian.
Replacing R by $R_{\mathcal{P}}$, we may assume that R is a noetherian semi local domain of dimension 1.
By [Ka; 74, 2-3, Ex. 21] the integral closure $\tilde{R}$ of R in its field of fractions is a semi local Dedekind domain, and hence a principal ideal domain [Ka; 74, 2-2, Ex. 21]. □

§2   Modules

Let R be a Dedekind domain with field of fractions K, and let Λ be an R–order in the separable K–algebra A.

We recall some facts about Λ–lattices. Let $_{\Lambda}\mathcal{M}^0$ denote the category of left Λ–lattices i.e. R–torsion free Λ–modules which are finitely generated over R.

2.1. Facts:.

(a) If S is a faithfully flat commutative ring extension of R, then

$$S \otimes_R Hom_\Lambda(M, N) \simeq Hom_{S \otimes_R \Lambda}(S \otimes_R M, S \otimes_R N)$$

for $M, N \in {}_\Lambda\mathcal{M}^0$.

(b) For $\wp \in max(R)$ and an R–lattice X, we denote by $X_\wp$ the localization of X at $\wp$. Given $M_\wp \in {}_{\Lambda_\wp}\mathcal{M}^0$, one for each $\wp \in max(R)$. Then there exists a left Λ–lattice M with

$$R_\wp \otimes_R M \simeq M(\wp)$$

if and only if

$$K \otimes_{R_\wp} M(\wp) \simeq V \text{ -}$$

for every $\wp$ and a fixed A–module V.

(c) $M_\wp \simeq N_\wp \Longleftrightarrow \hat{M}_\wp \simeq \hat{N}_\wp$ where "$\hat{X}$" denotes $\wp$–adic completion of X, $lim.proj.X/\wp^n \cdot X$.

(d) The Noether–Deuring theorem: Let $\hat{S}$ be a ring extension of $\hat{R}$ such that $\hat{S}$ is as $\hat{R}$–module of finite rank. Then for $\hat{M}, \hat{N} \in {}_{\hat{\Lambda}_\wp}\mathcal{M}^0$

$$\hat{S} \otimes_{\hat{R}} \hat{M} \simeq_{\hat{S} \otimes_R \Lambda} \hat{S} \otimes_{\hat{R}} \hat{M} \Longleftrightarrow \hat{M} \simeq_{\hat{\Lambda}} \hat{N}$$

(e) $M, N \in {}_\Lambda\mathcal{M}^0$ are said to lie in the same genus –notation $M \vee N$– iff $M_\wp \simeq N_\wp$ as $\Lambda_\wp$–lattices for every $\wp \in max(R)$. Here the localization can be replaced by the completion. Then

$$M \vee N \Longrightarrow M^{(n)} \simeq N^{(n)}$$

for some $n \in I\!N$. Moreover there exists a finite extension S of R with

$$S \otimes_R M \simeq S \otimes_R N.$$

(This holds in case $K = frac(R)$ is an algebraic number field.)

(f) Let P and Q be rank one locally free Λ–lattices. Then $KP \simeq KQ \simeq A$ and we may assume that we have an equality: $KP = KQ = A$, thus $\hat{P}_\wp = \hat{\Lambda}_\wp \alpha_\wp$ for an $\alpha_\wp \in U(A)$ and each $\wp$. Here $\alpha_\wp \in U(\hat{\Lambda}_\wp)$ for almost all $\wp$. $(\alpha) := \prod_\wp \alpha_\wp$ is called an idèle and we can multiply them component by component. Given an idèle (α) we can associate with it a module

$$\Lambda\alpha := P := \bigcap_\wp \Lambda\alpha_\wp \cap A.$$

The idèles in the set

$$(\prod_\wp U(\Lambda_\wp)\alpha_\wp)U(A)$$

all induce the same isomorphism class of lattices. Conversely this set gives all representatives of the isomorphism class of $\Lambda\alpha$ as long as they are subsets of A. With this technique we can prove that

$$\Lambda\alpha \oplus \Lambda\beta \simeq \Lambda \oplus \Lambda(\alpha \cdot \beta)$$

for idèles α, β.

We shall use the following general *notations*.

R is a noetherian integral domain of characteristic zero
$K = Frac(R)$ is the field of fractions of R .
G is a finite group.
$RG = \{\Sigma_{g \in G} r_g g : r_g \in R, g \in G\}$ is the group ring.
$|G| = \prod_{1 \le i \le \nu} p_i^{n_i}$ is the decomposition into different prime powers.
$\pi(G) = \{p_1, ..., p_\nu\}$ are the various prime divisors of $|G|$.
$\wp_{i,j}$ are the prime ideals of R above $p_i R$.
$R_{\pi(G)}$ is the localization of R at the set $R \setminus (\cup \wp_{i,j})$.
$U(RG)$ are the units in RG.
$\epsilon_G^R : RG \longrightarrow R$ is the augmentation map, induced by $g \longrightarrow 1$.
$I_R(G)$ is the kernel of the augmentation map.
$\tau_G^R : RG \longrightarrow RG$ induced by $g \longrightarrow g^{-1}$ is the antiinvolution associated with G in RG.
$V(RG)$ are the units in RG of augmentation 1.
$x(1)$ ist the coefficient of 1 in $x = \sum_{g \in G} r_g \cdot g \in RG$, i.e. $x(1) = r_1$.
$tr(x)$ is the trace of $x \in RG$ – note that this is just $|G| \cdot x(1)$.

II   Some notes on representation theory

§1   Blocks

[C-R2; 87, Chapter 8]
Let R be a complete Dedekind domain of characteristic zero with maximal ideal $\wp$ and assume that $\text{char}(R/\wp) = p > 0$, or let R be a field of characteristic $p > 0$, and let G be a finite group.

The indecomposable ring direct summands of RG are called the *blocks* of RG. The trivial RG-module R occurs with multiplicity one in RG, and hence there exists a *unique* block B_0, the *principal block*, which contains the trivial module.

1.1. Lemma. *Let G be a solvable group, then the principal block is*

$$B_0 = RG/O_{p'}(G),$$

where $O_{p'}(G)$ is the maximal normal subgroup of order prime to p .

PROOF OF LEMMA 1.1: The uniqueness of maximal normal p'−subgroups is easily seen. Let $N := O_{p'}(G)$. The exact sequence

$$0 \longrightarrow I_R(N) \cdot G \longrightarrow RG \longrightarrow RG/N \longrightarrow 0$$

is then two-sided split, since $\eta = (1/|N|) \cdot (\Sigma_{n \in N} n)$ is a central idempotent in RG with $RG \cdot \eta \simeq RG/N$. Surely RG/N contains the trivial module, and it remains to show that RG/N is indecomposable as ring. So we may assume, that $O_{p'}(G) = 1$. If P is a maximal normal p–subgroup of G, then $P > C_G(P)$, since $O_{p'}(G) = 1$. Let $I\!F = R/\wp$ be the residue field of R. Since RG is indecomposable as ring – $i.e.$ as $(G \times G)$–bimodule– iff $I\!FG$ is indecomposable as ring [3], it is enough to show that $I\!FG$ is indecomposable as ring. For this it suffices to show the following lemma:

This result on indecomposability of permutation modules, which was communicated to me by L. L. Scott during our collaboration, also gives a criterion for when the group ring consists only of the principal block:

1.2. Lemma. *Let $I\!F$ be a field of characteristic $p > 0$ and G a finite group, having a normal p–subgroup P with $C_G(P) \leq P$. Then the $(P \times G^{op})$–left module M, which is $I\!FG$ as vector space with the action $(p \times g) \cdot x = p \cdot x \cdot g$ is absolutely indecomposable - that is $M \simeq I\!F \uparrow_{\Delta(P)}^{P \times G^{op}}$, where*

$$\Delta(P) = \{(p, p^{-1}) | p \in P\}.$$

PROOF OF LEMMA 1.2: We show, that $End_{P \times G^{op}}(M)$ has a simple top, isomorphic to $I\!F$. We have

$$End_{P \times G^{op}}(M) \simeq I\!FG^P = \{x | x \in I\!FG : p \cdot x = x \cdot p\}.$$

A basis of $End_{P \times G^{op}}(M)$ is thus given by $\{K_P(g)\}$, the P–conjugacy class sums of G under conjugation action. Such a class sum $K_P(g)$ has length one iff $g \in C_G(P)$.

1.3. Claim. *The elements $K_P(g)$ for $g \notin C_G(P)$ lie in $rad(I\!FG)$.*

PROOF OF CLAIM 1.3: $K_P(g) = \Sigma_{i=1}^{t} p_i \cdot g \cdot p_i^{-1}$ for certain elements $p_i \in P$, where $t = p \cdot \tau$ is a multiple of p. We shall show that $K_P(g) \in I_{I\!F}(P) \cdot G$, the ideal generated by the augmentation ideal of P, a two sided ideal in the radical of $I\!FG$. However

$$K_P(g) = \sum_{i=1}^{t} p_i \cdot g \cdot p_i^{-1} = \sum_{i=1}^{t} p_i \cdot g \cdot p_i^{-1} - t \cdot g \in I_{I\!F}(P) \cdot G$$

iff

$$\sum_{i=1}^{t} (p_i \cdot g - g \cdot p_i) \in I_{I\!F}(P) \cdot G;$$

but,

$$p_i \cdot g - g \cdot p_i = (p_i - 1) \cdot g - g \cdot (p_i - 1) \in I_{I\!F}(P) \cdot G.$$

This proves the Claim 1.3.   □

---

[3]   This is the method of lifting idempotents (cf. [C-R1; 82, (18.1.iii)])

Since the elements $K_P(g)$ of length one contain the elements $g \in C_G(P) \subset P$, we have for these elements that $g = 1 + (g-1)$ is congruent to one modulo $I_{\mathbb{F}(P)}$. Thus the radical quotient of $End_{P \times G^{op}}(M)$ is $\mathbb{F}$ as claimed.
q.e.d. Lemma 1.2   □

Therefore also Lemma 1.1 is proved.   □

§2   Normalizers of p–Sylow subgroups of group bases

The following is an extension of a result of D. Coleman [Col; 64], which was communicated to me by L. L. Scott during our collaboration:

2.1. Lemma. *Let G be a finite group with a p–subgroup P and S an integral domain, in which p is not invertible. Then $N_{V(SG)}(P)/C_{V(SG)}(P) \simeq N_G(P)/C_G(P)$.* [4]

PROOF: By the arguments in Section I. 1.1 we may assume, that S is contained in a complete local principal ideal domain R, in which p is not invertible. Moreover, it suffices to prove the result for RG. Obviously, $N_G(P)/C_G(P)$ embeds into $N_{V(RG)}(P)/C_{V(RG)}(P)$. So let $v \in N_{V(RG)}(P)$ and denote by M the RG-bimodule RG with the action

$$g \bullet m \bullet h = v \cdot g \cdot v^{-1} \cdot m \cdot h, \ m \in M, g, h \in G \ .$$

Since v is a unit in RG, M is an RG-bimodule which is isomorphic to the natural bimodule RG, the isomorphism being given by $\sigma : m \longrightarrow v^{-1} \cdot m$ from M to RG. We now view M and RG as left RP–modules under the diagonal action:

$$p \bullet m = v \cdot p \cdot v^{-1} \cdot m \cdot p^{-1} \text{ for } m \in M, p \in P$$

and $p \bullet x = p \cdot x \cdot p^{-1}$ for $x \in RG$ and $p \in P$. The above isomorphism σ is then also an SP–module isomorphism. Since RG is a permutation module for P under the diagonal action, the same must be true for M. Since v normalizes P, a permutation basis for M as SP–module is given by $\{g \,|\, g \in G\}$. However, RG as RP–module has a P–trivial direct summand, namely R. Since P is a p–group and p is not invertible in R, a transitive permutation module RP/P_0 has a P–trivial direct summand iff $P_0 = P$. Thus the permutation basis $\{g \,|\, g \in G\}$ must have a fixed point g_0; i.e. for every $p \in P$ we have

$$v \cdot p \cdot v^{-1} \cdot g_0 \cdot p^{-1} = g_0 \ ; \text{ i.e. } v \cdot p \cdot v^{-1} = g_0 \cdot p \cdot g_0^{-1}.$$

Thus $v \in N_G(P) \cdot C_{V(RG)}(P)$ as claimed.   □

---

[4]
$$N_V(X) := \{v \in V : \ {}^v x \in X; \ \forall x \in X\}$$
is the normalizer in V of X and
$$C_V(X) := \{v \in V : \ {}^v x = x \ \forall x \in X\}$$
is the centralizer of X in V.

2.2. Corollary. *Let G be a finite group with $p-$subgroups P and Q, and let S be an integral domain, in which p is not invertible. Assume, that there is a unit $u \in U(SG)$ with $^uP = Q$, then there exists $g \in G$ with $^up = {}^gp$ for every $p \in P$. (I.e. if two subgroups P and Q are conjugate in $U(SG)$, then they are already conjugate in G.)*

PROOF: The argument is exactly the same as above, and we conclude

$$v \cdot p \cdot v^{-1} \cdot g_0 \cdot p^{-1} = g_0;$$

i.e.

$$v \cdot p \cdot v^{-1} = g_0 \cdot p \cdot g_0^{-1},$$

and so P and Q are $G-$conjugate, and v acts as u on P. □

§ 3   Cohomology rings

As general reference we give: [H-S; 70, Chapter 13] or [Br; 87, Chapter V,4]

Let R be a complete Dedekind domain of characteristic zero with maximal ideal $\wp$ and $\operatorname{char}(R/\wp) = p > 0$ or let R be a field of characteristic $p > 0$, and let G be a finite group.

The cohomology groups $H^i(G, M)$ for an $RG-$module M can be defined as

(1) $H^i(G, M) = Ext^i_{RG}(R, M)$ for $i > 0$,

and $H^0(G, M)$ consists of the $R-$module of $G-$fixed points in M. Equivalently, $H^0(G, M) \simeq Hom_{RG}(R, M)/proj(R, M)$, where $proj(R, M)$ are the homomorphisms from R to M which can be factored via a projective $RG-$module.

If M is indecomposable, then it lies in a unique block B of RG. If B is not the principal block, then $H^i(G, M) = 0$. If M lies in the principal block B_0, then $H^i(G, M) \simeq Ext^i_{B_0}(R, M)$.

3.1. The cup product. Let

$$H^*(G, R) = \bigoplus_{i \geq 0} H^i(G, R)$$

Then $H^*(G, R)$ is a graded $H^0(G, R) = R-$module. However, one can also make $H^*(G, R)$ into a ring, the *cohomology ring*:

We have a pairing:

(2) $H^i(G, R) \times H^j(G, R) \longrightarrow H^{i+j}(G, R),$

defined as follows: Because of equation (1) it is enough to define a pairing

(3) $Ext^i_{RG}(R, R) \times Ext^j_{RG}(R, R) \longrightarrow Ext^{i+j}_{RG}(R, R).$

If i or j is zero, then this is just the R-module structure of the cohomology groups. Interpreting the Ext-group as equivalence classes of module extensions, we compose the two extensions:

$$E_i : 0 \longrightarrow R \longrightarrow M_i \longrightarrow ... \longrightarrow M_1 \xrightarrow{\alpha} R \longrightarrow 0$$

and

$$E_j : 0 \longrightarrow R \xrightarrow{\beta} N_j \longrightarrow ... \longrightarrow N_1 \longrightarrow R \longrightarrow 0.$$

We form the exact sequence

$$E_{i+j} : 0 \to R \to M_i \to ... \to M_1 \xrightarrow{\alpha\beta} N_j \to ... \to N_1 \to R \to 0 \ .$$

The class of E_{i+j} in $H^{i+j}(G,R)$ is then the product of the class of E_i in $H^i(G,R)$ with the class of E_j in $H^j(G,R)$.

It is a tedious but straightforward calculation to show, that this gives a well defined pairing

$$H^i(G,R) \times H^j(G,R) \longrightarrow H^{i+j}(G,R) \ ,$$

which makes $H^*(G,R)$ into an associative ring. It is commutative in the graded sense, since one has for homogeneous elements

(4) $$x \cdot y = (-1)^{\delta(x) \cdot \delta(y)} y \cdot x \ ,$$

if $\delta(x)$ is the degree of x and x has degree i, iff $x \in H^i(G,R)$.

3.2. Cohmology variety. We note that $H^*(G,R)$ is a commutative ring, if R is a field of characteristic two. In general, $H^*(G,R)$ is only commutative in the graded sense, however, one can still talk about the prime spectrum with the Zariski topology.

More precisely, if for two homogeneous elements x and y the product $x \cdot y$ and $y \cdot x$ are different, then because of the above formula (4) both x and y must have odd degree. However, if x has odd degree i, then i^2 is odd, and thus $x \cdot x = (-1)^{i^2} x \cdot x = -x \cdot x$, and so $x^2 = 0$; $i.e.$, the elements of odd degree lie in the nil radical, $nil(H^*(G,R))$, of $H^*(G,R)$, and modulo the nil radical $H^*(G,R)$ is commutative. However the prime ideals all contain $nil(H^*(G,R))$, and thus are in bijection to the spectrum of the commutative ring $H^*(G,R)/nil(H^*(G,R))$. The spectrum of $H^*(G,R)$ with the Zariski topology is thus the same as the spectrum of $H^*(G,R)/nil(H^*(G,R))$ with the Zariski topology [Har; 77] and is called the *cohomology variety* of G . We denote this by $V(G,R)$.

§4   Profinite groups

As a general reference we give [Gr; 67, Chapter V]

Let

$$\ldots \longrightarrow G_i \xrightarrow{\varphi_i} G_{i-1} \longrightarrow \ldots \longrightarrow G_1$$

be a chain of maps of *finite groups* - more generally a projective system of finite groups [Bo; 60].

The *projective limit* of this chain

$$G = Lim.proj.\{G_i\} = \{(g_i)|g_i \in G_i : g_i\varphi_i = g_{i-1}\}$$

is a group, the multiplication being induced from the multiplication on the G_i . Moreover, G is a topological group, where the topology is induced from the discrete topology on the groups G_i . If the finite groups G_i are $p-$groups, then the projective limit G is said to be a *pro p−group*.

4.1. Examples.

(a) Let $\mathbb{Z}_i = \mathbb{Z}/(p^i \cdot \mathbb{Z})$. Then there are natural maps

$$Gl(n, \mathbb{Z}_i) \longrightarrow Gl(n, \mathbb{Z}_{i-1}),$$

and the projective limit of this chain is $Gl(n, \hat{\mathbb{Z}}_p)$.

(b) Let G be a finite group, then the unit group $U(\hat{\mathbb{Z}}_pG)$ is a profinite group.

(c) $Gl(n, \hat{\mathbb{Z}}_p)$ is not a pro $p−$group, but

$$1 + p \cdot Gl(n, \hat{\mathbb{Z}}_p) = Lim.proj.(1 + p \cdot Gl(n, \hat{\mathbb{Z}}_p)/1 + p^i \cdot Gl(n, \hat{\mathbb{Z}}_p)$$

is a pro $p−$group.

(d) If G is a finite $p−$group, then $V(\hat{\mathbb{Z}}_pG)$ is a pro $p−$group.

4.2. Continuous cohomology of profinite $p-$groups. Let

$$G = Lim.proj.\{G_i\}$$

be the projective limit of finite groups from the projective system

(5) $$\ldots \longrightarrow G_i \xrightarrow{\varphi_i} G_{i-1} \longrightarrow \ldots \longrightarrow G_1.$$

The homomorphisms $\varphi_i : G_i \longrightarrow G_{i-1}$ induces a map on cohomology

(6) $$\varphi_i^{j*} : H^j(G_{i-1}, M_{i-1}) \longrightarrow H^j(G_i, M_{i-1}) .$$

where M_{i-1} is a $G_{i-1}−$module, which is viewed via φ_i as $G_i−$module. Since (5) is a projective system, (6) is an injective system, and

(7) $$H_c^j(G, \hat{\mathbb{Z}}_p) = lim.inj.(H^j(G_i, \hat{\mathbb{Z}}_p)$$

is called the $j-th$ *continuous cohomology group* of G. (Because of the definition of the topology on G in terms of the discrete topology on G_i, these cohomology groups are exactly the quotient of the continuous j^{th} cocycles modulo the continuous j^{th} coboundaries from G to $\hat{\mathbb{Z}}_p$.)

III   The leading coefficient of units

The aim in this section is to prove the following

III.1. Theorem. *(Saksonov)[Sak; 71]. Let G be a finite group and R an integral domain of characteristic zero, in which no rational prime divisor of $|G|$ is invertible.*
If $u \in V(RG)$ is a unit of finite order n, then n is a divisor of $|G|$ and either $u(1) = 0$ or $u = u(1)$, where $u(1)$ is the coefficient of 1 in

$$u = \sum_{g \in G} r_g \cdot g \ .$$

In particular, if $u \in V(RG)$ is a non trivial unit of finite order, then $u(1) = 0$.

We shall first make some *historical remarks*.
The first who proved a theorem in this direction was G.Higman in his thesis from 1939. The result remained unpublished. Berman proved this theorem for $\mathbb{Z}$ in 1955, however, the result can also be found in Graham Higman's thesis from 1939 [San; 85]. More general versions were given by Glauberman in 1965 and Saksonov in 1968 [Gl; 65][Sak; 71].

Before we come to the PROOF let us draw some *consequences* of III.1. The hypotheses on R are as in Theorem III.1.

III.2. Corollary. *Let H be a finite subgroup of $V(RG)$, then $|H| \le |G|$, and the elements of H are R–linearly independent in RG.*
Moreover, if $|H| = |G|$, then $RG = RH$ as augmented algebras.

PROOF OF COROLLARY III.2: Assume that

$$\sum_{h \in H} r_h \cdot h = 0$$

with $r_{h_0} \ne 0$. Multiplying the above equation by h_0^{-1}, we may assume that $r_1 \ne 0$. By Theorem III.1 we have for every $h \in H$ viewed as an element $h \in RG$, the trace, $tr(h) = 0$ of h is zero, unless $h = 1$. Thus

$$0 = tr(\sum_{h \in H} r_h \cdot h) = \sum_{h \in H} r_h tr(h) = r_1 \cdot |G| \ ,$$

a contradiction.

If now $|H| = |G|$, then $KG = KH$ because of the above result – $K = Frac(R)$. Moreover, we surely have $RH \subset RG$. Let

$$g = \sum_{h \in H} k_h \cdot h \text{ with } k_h \in K \ .$$

Then $g \cdot h^{-1} \in RG$ and hence it has trace in R .
On the other hand, the trace can be computed in $KH = KG$ and then

$$tr(g \cdot h^{-1}) = k_h \ ;$$

thus $k_h \in R$ and so $RG = RH$.

This completes the proof of Corollary III.2   □

III.3. Corollary. *Let H be a finite subgroup of $V(RG)$, then KG as left KH-module is free. In particular, $|H| \, | \, |G|$*

III.4. Note. For RG this is false. We construct a counterexample in section VII Lemma 3.3.

PROOF OF COROLLARY III.3: Let $\tilde{K}$ be an extension field of K, which is a splitting field for G^5. Because of the Noether-Deuring theorem[6] [C-R1; 82, §6, Exercise 6], it is enough to assume that $\tilde{K}G$ is $\tilde{K}H$-free. However, the trace of H on $\tilde{K}G$ is the character of $\tilde{K}G$, and by Theorem III.1 this is the same as the character of a free module. But $\tilde{K}H$-modules are determined by their character.
q.e.d.   □

III.5. Note. One actually would like to have a better result than Theorem III.1: Assume that for the rational prime P the ideal $p \cdot R \neq R$. If $u \in V(RG)$ is an element of order p, is then still $u(1) = 0$?
This however is false. We construct a counterexample in XIV.1.

We now come to the

PROOF OF THEOREM III.1: We first need an *observation on idempotents:*

III.6. Proposition. *Let K be a field of characteristic zero, and let*

$$e = \sum_{g \in G} e(g) \cdot g$$

be a non trivial idempotent in KG [7]. Then

$$e(1) \in \mathbb{Q},$$

and, moreover,

$$0 < e(1) < 1.$$

---

5   i.e. $End_{\tilde{K}G}(V) = \tilde{K}$ for every simple RG–module.
6   Let A be a finite dimensional K–algebra over the field K with extension field E and let M and N be finitely generated left A-modules. Then

$$E \otimes_K M \simeq_{E \otimes_K A} E \otimes_K N \Longleftrightarrow M \simeq_A N.$$

7   i.e. $0 \neq e \neq 1$

PROOF OF PROPOSITION III.6: We shall compute the trace of e in two ways: With respect to the basis G in KG we have

$$tr(e) = e(1) \cdot |G| .$$

On the other hand,

$$KG = KG \cdot e \; \oplus \; KG \cdot (1 - e)$$

and so
(8) $$tr(e) = dim_K(KG \cdot e) =: n_e .$$

However, e is non trivial, and so

$$0 < n_e < |G| ;$$

but then

$$0 < e(1) = \frac{n_e}{|G|} < 1 .$$

q.e.d.   □

III.7. Corollary. *Let R be as in Theorem III.1. Then RG has no non trivial idempotents.*

PROOF OF COROLLARY III.7: Let e be a non trivial idempotent in RG. Because of Proposition 1.1 in section I we may assume that

  (a) R is a principal ideal domain,
  (b) $R \cap \mathbb{Q} = \mathbb{Z}_{\pi(G)}$.

Because of Proposition III.6 in section I we conclude, that (cf. the equation (8))

$$e(1) = \frac{n_e}{|G|} \in \mathbb{Z}_{\pi(G)} = R \cap \mathbb{Q},$$

and thus

$$n_e \in |G| \cdot \mathbb{Z}_{\pi(G)}.$$

Noting that $n_e \in \mathbb{Z}$ and $0 < n_e < |G|$, we reach a contradiction, by looking at the prime factorization of n_e and $|G|$. q.e.d.   □

In the same spirit we can also prove the

III.8. Claim. *Let H be a finite subgroup of the augmented units of RG. Then RH is a pure sublattice in RG.*[8]

PROOF OF CLAIM III.8: Let

$$x = \sum_{g \in G} r_g g \in RG \setminus RH$$

---

8   i.e. a sublattice N of M such that M/N is R-torsion free.

with $tx \in RH$ for some $t \in R$. Then

$$x = \sum_{h \in H} k_h h$$

with $k_h \in K$ and $tk_h \in R$. Therefore

$$t \cdot tr(x) = \sum_{h \in H} tr(tk_h h) = \sum_{h \in H} t \cdot tr(k_h h).$$

By Theorem III.1 we get

$$t \cdot r_1 \cdot |G| = t \cdot |G| \cdot k_1 \cdot 1$$

and thus $k_1 \in R$. The same arguments applied to $h^{-1}x$ yields $k_h \in R$ and hence $k_h \in R$ for all $h \in H$. $\qquad\square$

We are now in the position to PROVE THEOREM III.1 [9], which we recall for the reader's convenience:

Theorem. *(Saksonov)[Sak; 71] Let G be a finite group and R an integral domain of characteristic zero, in which no rational prime divisor of $|G|$ is invertible.*
If $u \in V(RG)$ is a unit of finite order n, then n is a divisor of $|G|$ and either $u(1) = 0$ or $u = u(1)$, where $u(1)$ is the coefficient of 1 in

$$u = \sum_{g \in G} r_g \cdot g \ .$$

In particular, if $u \in V(RG)$ is a non trivial unit of finite order, then $u(1) = 0$.

Invoking Proposition 1.1 of section I we may assume — if necessary enlarge R:

(a) R is a principal ideal domain,
(b) $R \cap \mathbb{Q} = \mathbb{Z}_{\pi(G)}$,
(c) $K = frac(R)$ contains a normal algebraic number field K_0 , which is a splitting field for G and which is invariant under complex conjugation,
(d) We put $R_0 = alg.int.(K_0) \subset R$,
(e) $\zeta_n \in R$, where ζ_n is a primitive n^{th} root of unity.

Let $R_0 C_n$ be the group ring of $C_n = <c : c^n = 1>$, the cyclic group of order n, over R_0 . Then in $K_0 C_n$ the element c is represented as

$$c = \sum_{1 \le j \le n} \zeta_n^i \cdot \epsilon_j \ ,$$

---

[9]   The original proof was improved very much by cand. math. Ulrike Villinger (University Stuttgart) while preparing a seminar in spring 1991 at the University of Stuttgart and her version is presented here.

where $\{\epsilon_j\}_{1 \leq j \leq u}$ is a complete set of orthogonal primitive idempotents in $K_0 C_n$, numbered appropriately.

Since $u \in V(RG)$ has order n, we get a natural unitary ring homomorphism

$$\varphi : K_0 C_n \longrightarrow KG$$

induced by $\varphi(c) = u$, and the image u is represented as

$$(9) \qquad u = \sum_{1 \leq i \leq n} \zeta_n^i \cdot e_i \text{ , with } u^k = \sum_{1 \leq i \leq n} \zeta_n^{i \cdot k} \cdot e_i \text{ ,}$$

where $\{e_i\}$ are orthogonal but not necessarily nontrivial[10] nor primitive idempotents in KG defined by $\varphi(\varepsilon_i) =: e_i$. Moreover,

$$1 = \sum_{1 \leq i \leq n} e_i \text{ .}$$

Since K_0 is a splitting field for G, the idempotents $\{e_i\}$ lie already in $K_0 G$. The elements $\{\zeta_n^i\}$ also lie in K_0 . And so we conclude

$$(10) \qquad u \in R_1 G, \text{ where } R_1 = R \cap K_0 \text{ .}$$

Thus we may assume that

$$(11) \qquad\qquad K \text{ is an algebraic number field.}$$

We now look at the coefficient of 1 in u:

$$(12) \qquad u(1) = \sum_{1 \leq i \leq n} \zeta_n^i \cdot e_i(1) \text{ and } u^k(1) = \sum_{1 \leq i \leq n} \zeta_n^{i \cdot k} \cdot e_i(1),$$

because of Equation 9.

If all the e_i are trivial, then $u \in K$ and the theorem is established. If not, then $\{e_i\}_{1 \leq i \leq n_u}$ are w.l.o.g. the non trivial idempotents.

We now invoke Proposition III.6:

$$(13) \qquad u^k(1) = \sum_{1 \leq i \leq n_u} \zeta_n^{k(i) \cdot k} \cdot \frac{n_{e_i}}{|G|} \text{ with } \sum_{1 \leq i \leq n_u} n_{e_i} = |G|$$

for exponents $k(i)$ depending on the renumbering of the $\{e_i\}$ above. This shows in particular, that

$$u(1) \in \mathbb{Q}(\zeta_n) \cap R \text{ .}$$

Let now

$$\mathcal{G} = Gal(\mathbb{Q}(\zeta_n) : \mathbb{Q})$$

be the Galois group of the cyclotomic field $\mathbb{Q}(\zeta_n)$, and note that $\mathbb{Q}(\zeta_n) \subset K$. Since K is Galois, $\sigma \in \mathcal{G}$ extends to an automorphism of K, also denoted by

---

10   i.e. not equal to 0 or 1

σ. The Equation 12 shows that all the Galois conjugates under $\sigma \in \mathcal{G}$ of $u(1)$ are of the form $u(1)^k$.

III.9. Claim. *Denote by Nr the norm from $\mathbb{Q}(\zeta_n)$ to $\mathbb{Q}^{11}$. Then*

$$Nr(u(1)) \in \mathbb{Z}.$$

PROOF OF CLAIM III.9: We have

$$|G| \cdot u(1) \in \mathbb{Z}[\zeta_n],$$

We now take the norm from $\mathbb{Q}(\zeta_n)$ to $\mathbb{Q}$ – note that $Nr(u(1)) \in R \cap \mathbb{Q} = \mathbb{Z}_{\pi(G)}$, and obtain

$$
\begin{aligned}
(14) \qquad Nr(|G| \cdot u(1)) \ &= \ |G|^{\mu(n)} \cdot Nr(u(1)) \\
&\in \ \mathbb{Z} \cap |G|^{\mu(n)} \cdot \mathbb{Z}_{\pi(G)} = |G|^{\mu(n)} \cdot \mathbb{Z} \ ,
\end{aligned}
$$

were $\mu(n)$ denotes the dimension of the cyclotomic field $\mathbb{Q}(\zeta_n)$ over $\mathbb{Q}$. Hence $Nr(u(1)) \in \mathbb{Z}$. q.e.d. Claim III.9 □

Note that the same argument applies to $u^k(1)$.

Let $x \longrightarrow \overline{x}$ denote the complex conjugation. Then the triangle inequality shows — using Equation (13):

$$(15) \qquad u(1) \cdot \overline{u(1)} = |u(1)|^2 \ \leq \ (\sum_{1 \leq i \leq n_u} |\zeta_n^{k(i)}| \cdot \frac{n_{e_i}}{|G|})^2 = 1.$$

All Galois conjugates of $u(1)$ have absolute value less or equal than 1. Complex conjugation is a Galois–automorphism and hence $Nr(u(1)) \leq 1$. Since this number is a rational integer, we see from III.9 that the norm is equal to 1 or to 0. If the latter happens, then $u(1) = 0$ and the theorem is proved. If not, $|u(1)|^2 = 1$ and we conclude:

(16) $\quad u(1)$ and all its conjugates under $\sigma \in \mathcal{G}$ have absolute value 1 .

Since
$$(17) \qquad u(1) = \sum_{1 \leq i \leq n_u} \zeta_n^{k(i)} \cdot \frac{n_{e_i}}{|G|} \ , \ \text{with} \ \sum_{1 \leq i \leq n_u} \frac{n_{e_i}}{|G|} = 1 \ ,$$

the element $u(1)$ lies in the convex hull of the set $\{\zeta_n^{k(i)}\}$. But $u(1)$ has absolute value 1 and so $u(1)$ is a root of unity. This implies

$$(18) \qquad \zeta_n^{k(i)} = \zeta(0) \ \text{for every} \ 1 \leq i \leq n_u \ .$$

But then $n_u = 1$ and we reach a contradiction. q.e.d. Theorem III.1 □

We may also conclude from this discussion that the elements of a group bases are linearly independent.

---

11   i.e. $Nr(x) = \prod_{\sigma \in \mathcal{G}} {}^{\sigma}x$

IV Class sum correspondence

A class sum $K_g^G = \sum_{x \in G/C_G(g)} {}^x g$ is the sum in RG of all elements that are conjugate in G to $g \in G$, these elements form a basis of the center $\mathcal{Z}(RG)$ of RG.

We shall prove here the result that the class sums of elements in G are determined by RG; a result with many interesting consequences.

IV.1. Theorem. *(Glauberman, Passman, Saksonov, Roggenkamp–Scott)*
[Gl; 65, Pa; 65, Sak; 71, R-S; 87 1] Let R be an integral domain of characteristic zero and let G and H be finite groups with

$$\mathbb{C}G \simeq \mathbb{C}H.$$

Assume that no rational prime divisor of $|G|$ is a unit in R.
If we have an equality of the centers of RG and RH,

$$\mathcal{Z}(RG) \;=\; \mathcal{Z}(RH)$$

as augmented algebras — note that the augmentation of RG induces an augmentation on $\mathcal{Z}(RG)$ —, then there exists a bijection[12]

$$\kappa \;:\; G \longrightarrow H,$$

such that in $\mathcal{Z}(RG)$ one has for the class sums

$$K_g^G \;=\; K_{\kappa(g)}^H \;;$$

i.e. the class sums of G and H coincide in $\mathcal{Z}(RG)$. We shall call this a bijection inducing the class sum correspondence.

IV.2. Remark. In the literature, the assumption is as far as we know always that
$$RG \;=\; RH$$
as augmented algebras. However, a careful analysis of the proof shows, that one actually only needs that

$$\mathbb{C}G \;=\; \mathbb{C}H \text{ and } \mathcal{Z}(RG) \;=\; \mathcal{Z}(RH).$$

PROOF OF THEOREM IV.1: We have to recall some well known results from ordinary representation theory:

IV.3. Note.

---
12   not necessarily a group homomorphism

(a) Orthogonality relations:

Let $\{\chi_i\}_{1 \le i \le s}$ be the different complex characters of G – s is the the number of conjugacy classes of G. We then have the following *scalar product*:

$$(x, y)_0 \; := \; \sum_{1 \le i \le s} \chi_i(x) \cdot \overline{\chi_i(y)} \; = \; \delta_{K_x^G, K_y^G} \frac{|G|}{|K_x^G|} = \delta_{K_x^G, K_y^G} |C_G(x)|,$$

where '$-$' is complex conjugation, $\delta_{K_x^G, K_y^G}$ is the Kronecker symbol and x and y are elements in G (cf. [C-R1; 82]).

(b) The central primitive idempotents in $\mathbb{C}G$ are given by:

$$e_i \; = \; \frac{\chi_i(1)}{|G|} \cdot \sum_{x \in G} \overline{\chi_i(x)} \cdot x \; = \; \frac{\chi_i(1)}{|G|} \cdot \sum_{K_x} \overline{\chi_i(x)} \cdot K_x,$$

cf. [C-R; 62].

(c) It should be noted that both, the class sums and the central primitive idempotents, form a *basis for the center of* $\mathbb{C}G$.

Using (a) and (b) one obtains

$$K_x \; = \; |K_x| \cdot \sum_{1 \le i \le s} \chi_i(x) \cdot \frac{e_i}{\chi_i(1)}.$$

Again by extending R according to Proposition 1.1 in section I, if necessary, we may assume:

IV.4. Assumptions.

(a) R is a principal ideal domain,

(b) $R \cap \mathbb{Q} = \mathbb{Z}_{\pi(G)}$,

(c) $K = frac(R)$ contains a normal algebraic number field K_0, which is a splitting field for G, and which is invariant under complex conjugation,

(d) we put $R_0 = alg.int.(K_0) \subset R$,

(e) $\zeta_n \in R$, where ζ_n is a primitive n^{th} root of unity and $n = |G|$,

(f) $\mathbb{C}G = \mathbb{C}H$ and $\mathcal{Z}(RG) = \mathcal{Z}(RH)$.

IV.5. Proposition. *Let* $\{g_i\}_{1 \le i \le s}$ *and* $\{h_j\}_{1 \le j \le s}$ *be sets of representatives of the conjugacy classes in* G *and* H *resp.* [13] *For a class sum* $K_{h_j}^H$ *in* RH *we write*

$$K_{h_j}^H \; = \; \sum_{1 \le i \le s} d_{h_j, g_i} \cdot K_{g_i}^G \in \mathcal{Z}(RG).$$

Then $d_{h_j, g_i} \in \mathbb{Z}[\zeta_n]$.

---

[13]   Note that the number of conjugacy classes of G and H coincide, since the class sums form a basis for the center of the group ring.

PROOF OF PROPOSITION IV.5: Because of the note IV.3 we get

$$(19) \quad |K^H_{h_j}| \cdot \sum_{1 \le k \le s} \chi^H_k(h_j) \cdot \frac{e_k}{\chi^H_k(1)} = \sum_{1 \le i \le s} \sum_{1 \le l \le s} d_{h_j,g_i} \cdot |K^G_{g_i}| \cdot \chi^G_l(g_i) \cdot \frac{e_l}{\chi^G_l(1)}.$$

Here we have written χ^G_i and χ^H_i for the irreducible characters of G and H resp., numbered according to the identification $\mathbb{C}G = \mathbb{C}H$, moreover, the central primitive idempotents of $\mathbb{C}G$ and $\mathbb{C}H$ coincide.

Comparing the coefficients in Equation (19) – observing, that the $\{e_k\}$ are linearly independent and $\chi^H_i(1) = \chi^G_i(1)$ since $\mathbb{C}G = \mathbb{C}H$ – we obtain:

$$(20) \quad |K^H_{h_j}| \cdot \chi^H_k(h_j) = \sum_{1 \le i \le s} d_{h_j,g_i} \cdot |K^G_{g_i}| \cdot \chi^G_k(g_i).$$

Forming the scalar product defined in Note IV.3, we get

$$(21) \quad |K^H_{h_j}| \cdot \sum_{1 \le k \le s} \chi^H_k(h_j) \cdot \overline{\chi^G_k(g_0)} =$$

$$= \sum_{1 \le i \le s} \sum_{1 \le k \le s} |K^G_{g_i}| \cdot d_{h_j,g_i} \cdot \chi^G_k(g_i) \cdot \overline{\chi^G_k(g_0)} = |G| \cdot d_{h_j,g_0}.$$

This holds for every g_0. Since

$$\frac{|G|}{|K_{h_j}|} = \frac{|H|}{|K_{h_j}|} = |C_H(h_j)|$$

is the order of the centralizer in H of h_j, we have:

$$(22) \quad d_{h_j,g_i} \cdot |C_H(h_j)| = \sum_{1 \le k \le s} \chi^H_k(h_j) \cdot \overline{\chi^G_k(g_i)}.$$

This holds for all $1 \le i, j \le s$. Since the right hand side is an algebraic integer in $\mathbb{Q}(\zeta_n)$ – note that the character values of G and H are sums of nth roots of unity – we conclude

$$(23) \quad d_{h_j,g_i} \cdot |C_H(h_j)| \in \mathbb{Z}[\zeta_n] \text{ and } d_{h_j,g_i} \in \mathbb{Q}(\zeta_n).$$

We now apply similar arguments as in section III: We take the norm from $\mathbb{Q}(\zeta_n)$ down to $\mathbb{Q}$ and note that the Galois conjugates with respect to $Gal(\mathbb{Q}(\zeta_n) : \mathbb{Q})$ of d_{h_j,g_i} also lie in R because of our conditions on R (Assumption IV.4; R is the maximal $R \cap \mathbb{Q}$–order in K. Since the algebraic numbers in a field form a maximal order and tensoring with a semi localization preserves this property [Re; 75, (11.1)].). We then have

$$(24) \quad \begin{aligned} Nr(d_{h_j,g_i} \cdot |C_H(h_j)|) &= |C_H(h_j)|^{\mu(n)} \cdot Nr(d_{h_j,g_i}) \\ &\in |C_H(h_j)|^{\mu(n)} \cdot (\mathbb{Q} \cap R) \\ &= |C_H(h_j)|^{\mu(n)} \cdot \mathbb{Z}_{\pi(G)} \end{aligned}$$

where $\mu(n)$ is the degree of $|\mathbb{Q}(\zeta_n) : \mathbb{Q}|$.

However, $|C_H(h_j)|$ is a divisor of $|H|$, and so we conclude by [Has; 49, p.238] that

(25) $Nr(d_{h_j,g_i}) \in \mathbb{Z}$ and so $d_{h_j,g_i} \in \mathbb{Z}[\zeta_n]$.

q.e.d. Proposition IV.5   □

IV.6. Note. Our conditions on R guarantee that

$$Nr(d_{h_j,g_i} \cdot |C_H(h_j)|) \in |C_H(h_j)|^{\mu(n)} \cdot \mathbb{Z}_{\pi(G)},$$

and we need, that *no rational prime divisor of* $|C_H(h_j)|, 1 \leq j \leq s$, *is a unit in* $R \cap \mathbb{Q}$. This condition is a little bit weaker than our condition on R. (However, one can not do with a condition like the following: If h_j is a p-element, then p is not invertible in R.)

Because of the above result, we shall *assume from now on that*

$$R = \mathbb{Z}[\zeta_n].$$

Let us recall the scalar product defined in IV.3: Let

$$a = \sum_{1 \leq i \leq s} a_i \cdot K_{g_i}^G \text{ and } b = \sum_{1 \leq i \leq s} b_i \cdot K_{g_i}^G$$

be elements in $\mathcal{Z}(RG)$. Then we have defined a scalar product

$$(a,b)_0 = \sum_{1 \leq i \leq s} a_i \cdot \bar{b}_i \cdot |K_i^G| \cdot |G|.$$

IV.7. Note. The above scalar product is defined in terms of the group basis G in $\mathbb{Q}(\zeta_n)G$ depending on the characters – i.e. traces of matrices – and the orthogonality relations. If now $KG = KH$, then $\mathcal{Z}(\mathbb{Q}(\zeta_n)G) = \mathcal{Z}(\mathbb{Q}(\zeta_n)H)$, since $\mathbb{Q}(\zeta_n)$ is a splitting field for G and H. Moreover the orthogonality relations IV.3 show – after normalizing – that both bases $\{K_{g_i}\}_{1 \leq i \leq s}$ and $\{K_{h_j}\}_{1 \leq j \leq s}$ are orthonormal bases for $\mathcal{Z}(\mathbb{Q}(\zeta_n)G)$.

We now define a new scalar product in the following way:

IV.8. Definition.

$$(a,b) = \frac{1}{|G|} \cdot \sum_{\sigma \in Gal(Q(\zeta_n):Q)} (a,b)_0^\sigma$$

A simple calculation then shows

(26) $(a,b) = \sum_{1 \leq j \leq s} \sum_{\sigma \in Gal(Q(\zeta_n):Q)} a_j^\sigma \cdot \bar{b}_j^\sigma \cdot |K_{g_j}^G|.$

IV.9. Definition. Let

$$\mathcal{B} = \{\beta_i\}_{1 \leq i \leq s} \text{ be a basis of } \mathcal{Z}(RG)$$

with

$$\beta_j = \sum_{1 \le i \le s} b_{j,i} \cdot K_{g_i}^G \text{ and } b_{j,i} \in \mathbb{Z}[\zeta_n].$$

Then the *weight* [Pa; 65] of $\mathcal{B}$ is defined as

$$
\begin{aligned}
w(\mathcal{B}) &= \sum_{1 \le i \le s} (\beta_i, \beta_i) \\
&= \sum_{1 \le i \le s} \sum_{1 \le j \le s} \sum_{\sigma \in Gal(Q(\zeta_n):Q)} b_{i,j}^\sigma \cdot \overline{b}_{i,j}^\sigma \cdot |K_{g_j}^G| \\
&= \sum_{1 \le i \le s} \sum_{1 \le j \le s} \sum_{\sigma \in Gal(Q(\zeta_n):Q)} |b_{i,j}^\sigma|^2 \cdot |K_{g_j}^G|.
\end{aligned}
$$

In this definition we have already used that the complex conjugation commutes with $\sigma \in Gal(\mathbb{Q}(\zeta_n) : \mathbb{Q})$.

For the basis $\mathcal{B}_G = \{K_{g_i}^G\}_{1 \le i \le s}$ the weight is computed as:

$$(27) \quad w(\mathcal{B}_G) = \sum_{1 \le i \le s} \sum_{\sigma \in Gal(Q(\zeta_n):Q)} |K_{g_i}^G| = |Gal(\mathbb{Q}(\zeta_n) : \mathbb{Q})| \cdot |G|.$$

Because of the Note IV.7 we conclude that also for the basis $\mathcal{B}_H = \{K_{h_j}^H\}_{1 \le j \le s}$ the relation

$$(28) \qquad w(\mathcal{B}_H) = |Gal(\mathbb{Q}(\zeta_n) : \mathbb{Q})| \cdot |G|$$

holds. We choose the numbering in the basis $\mathcal{B}$ such that $b_{i,i} \ne 0$. Then we get the following inequality:

$$
\begin{aligned}
w(\mathcal{B}) &= \sum_{1 \le i \le s} \sum_{1 \le j \le s} \sum_{\sigma \in Gal(Q(\zeta_n):Q)} |b_{i,j}^\sigma|^2 \cdot |K_{g_i}| \\
&\ge \sum_{1 \le i \le s} \sum_{\sigma \in Gal(Q(\zeta_n):Q)} |b_{i,i}^\sigma|^2 \cdot |K_{g_i}| \\
&\ge |Gal(\mathbb{Q}(\zeta_n) : \mathbb{Q})| \cdot |G| = w(\mathcal{B}_G).
\end{aligned}
$$

(29)

We have equality in the above inequality (29) if and only if $b_{i,j} = 0$ for $i \ne j$ and $b_{i,i}$ is a root of unity. This follows from the fact, proved by elementary analysis, that if a product $\prod_{i=1}^n \alpha_i$ of real numbers is equal to 1, then $\sum_i \alpha_i^2 \ge n$ – the rest of the arguments are the same as at the end of Section III.

Thus we have proved

IV.10. Proposition. *Let*

$$\mathcal{B} = \{\beta_i\}_{1 \le i \le s} \text{ be a basis of } \mathcal{Z}(RG)$$

with

$$\beta_j = \sum_{1 \leq i \leq s} b_{j,i} \cdot K^G_{g_i} \text{ and } b_{i,j} \in \mathbb{Z}[\zeta_n].$$

Then $w(\mathcal{B}) = w(\mathcal{B}_G)$ if and only if

$$b_{i,j} = 0 \text{ for } i \neq j \text{ and } b_{i,i} \text{ is an } n^{th} \text{ root of unity.}$$

It remains to show that the Basis $\mathcal{B}_H$ has minimal weight. But this is exactly the statement of Equation (28). Thus in the equation

$$(30) \qquad\qquad K^H_{h_j} = \sum_{1 \leq i \leq s} d_{h_j,g_i} \cdot K^G_{g_i}$$

we may conclude with Proposition IV.10 that

$$K^H_{h_i} = \zeta_i \cdot K^G_{g_i},$$

where ζ_i is an nth root of unity.
However, $\mathcal{Z}(RG) = \mathcal{Z}(RH)$ as augmented rings – induced from the augmentation of RG and RH – and so $\zeta_i = 1, 1 \leq i \leq s$.

> *This completes the proof of Theorem IV.1*   □

An immediate consequence of Theorem IV.1 is the following: Let $RG = RH$ as augmented algebras. If $c \in H$ is central, then $c \in G$. This requires that c is part of a group basis.

We shall show next that the assumption that c is part of a group bases H is not necessary.

IV.11. Theorem. *(Berman, Higman, Jackson)[Hig; 39, Ja; 69, Ber; 53] Let R be an integral domain of characteristic zero and G a finite group. Assume that no rational prime divisor of $|G|$ is a unit in R. If $c \in V(RG)$ is a central unit of finite order, then $c \in G$.*

PROOF OF THEOREM IV.11: If

$$u = \sum_{g \in G} u(g) \cdot g$$

is a central unit of finite order with $u(g_0) \neq 0$. Then $u \cdot g_0^{-1}$ is also a unit of finite order. But according to Theorem III.1 we see that $u \cdot g_0^{-1} = g_1 \in G$ and so $u \in G$.

q.e.d.   □

V   More on the class sum correspondence

by W.Kimmerle

The object of this section is the proof of further properties of the class sum correspondence. In particular we study the so-called powermap, i.e the behaviour of corresponding class sums under powers, and collect properties of a finite group determined by its character table. The consequences with respect to the isomorphism problem are the content of the following summarizing result.

V.1. Theorem. *Let R be an integral domain of characteristic zero and let G be a finite group. Assume that no prime divisor of $|G|$ is invertible in R. Let H be a group basis of RG, then there exists a bijection $\sigma : G \to H$ with the following properties:*

(a) *g and $\sigma(g)$ have the same class sums.*

(b) *g and $\sigma(g)$ have the same order and its conjugacy classes have the same length.*

(c) *$\sigma(g^n)$ is conjugate to $\sigma(g)^n$ for each $n \in \mathbf{N}$. σ can be chosen such that $\sigma(g^p) = \sigma(g)^p$ for each p-element $g \in G$.*

(d) *Let K be a field containing R. Let $\chi_1, \ldots, \chi_s$ be the irreducible K-characters of G. Then $\chi_i(g) = \chi_i(\sigma(g))$. In particular, choosing K big enough, it follows that G and H have the same ordinary character table.*

(e) *If N is a normal subgroup of G, then $\sigma(N)$ is a normal subgroup of H with the same order as N. σ actually induces a lattice isomorphism between the lattices of normal subgroups of G and of H.*
   The map $\bar{\sigma} : G/N \to H/\sigma(N)$ defined by $gN \to \sigma(g)\sigma(N)$ has the same properties for G/N and $H/\sigma(N)$ as σ.

(f) *[K-L-S-T; 90] If N is a minimal normal subgroup of G, then $\sigma(N)$ is isomorphic to N. This yields inductively, that G and H have the same set of chief series. More precisely, if*

$$1 = K_0 < K_1 < ... < K_n = G$$

is a chief series of G, then H has a chief series

$$1 = L_0 < L_1 < ... < L_n = H$$

such that

$$K_i/K_{i-1} \cong L_i/L_{i-1} \text{ for each } i \in \{1, 2, ..., n\}.$$

V.2. Comments. a) Part (a) has been proved in section IV.1.

b) An irreducible K-character χ of G is the trace with respect to K afforded by a simple KG-module. Extend χ K-linearly to KG. Then χ defines in a natural way an irreducible K-character on each group basis of KG which we denote for the sake of simplicity again by χ.

c) In [Sak; 66] two finite groups G and H are called a Brauer pair, if they satisfy the following property:

There are bijections $\chi \to \chi^*$ and $C \to C^*$ of the sets of the ordinary irreducible characters and conjugacy classes, respectively, of G and H, such that $\chi^*(C^*) = \chi(K)$ for all χ, C, and the power map on the conjugacy classes is compatible with $*$.

R.Brauer raised the question [Bra; 63, p.138] as to whether such G and H are isomorphic. E.Dade [Da; 64,2] has shown that G and H need not be isomorphic, if they are a Brauer pair.

Note that parts (a) to (d) of Theorem V.1 imply that G and H form a Brauer pair, if they have isomorphic integral group rings.

We see later that parts (e) and (f) of the Theorem already can be obtained under the weaker hypothesis that G and H form a Brauer pair or even weaker that G and H have the same ordinary character table.

d) The first goal in this section is to establish parts (b), (c) and (d) of the Theorem. Character tables are discussed in paragraph 2.

e) Part (e) is usually called the **normal subgroup correspondence**. One sees easily that part (e) has the following consequence. If M and N are corresponding normal subgroups of G and of H resp., then the augmentation ideals $I(M) \cdot RG$ and $I(N) \cdot RG$ coincide.

§1   Power map

In this section we do not assume that G is finite.

1.1. Proposition. *Let R be a ring and let H be a group basis of RG. Assume that there exists a class sum correspondence $\sigma : G \to H$ (note that a class sum correspondence in the case of infinite groups can only exist, if all conjugacy classes have finite length). Then $K^G_{\sigma(g)^n} = K^G_{\sigma(g^n)}$ provided for each $p \in \pi(n)$ one has $pR \neq R$.*

PROOF: For an element $x = \sum_{g \in G} r_g$ of RG and a given conjugacy class $C(g)$ we define

$$tr_{C(g)}(x) = \sum_{y \in C(g)} r_y.$$

Consider the free R-module (RG, RG) defined by

$$(RG, RG) = \{x \in RG; tr_{C(g)}(x) = 0 \text{ for all } g \in G\}.$$

Then (RG, RG) is the R-module of additive commutators. An easy calculation shows:

$$
\begin{aligned}
(RG, RG) &= {}_R < ab - ba; a, b \in RG > \\
&= {}_R < fg - gf; f, g \in G > \\
&= {}_R < dh - hd; d, h \in H >= (RH, RH).
\end{aligned}
$$

1.2. Claim. $\sigma(g) = g + x$ *with* $x \in (RG, RG)$.

PROOF OF THE CLAIM: Certainly we get $K_{\sigma(g)} = l_{\sigma(g)} \cdot \sigma(g) + x_1$ with $x_1 \in (RH, RH)$ and $l_{\sigma(g)}$ denotes the length of the conjugacy class of $\sigma(g)$. Analogously we obtain $K_g = l_g \cdot g + x_2$ with $x_2 \in (RG, RG)$. The augmentation map shows that $l_g = l_{\sigma(g)}$. Hence $l_g \cdot (\sigma(g) - g) \in (RG, RG)$ and the latter happens if, and only if, $\sigma(g) - g \in (RG, RG)$. $\qquad\square$

It suffices for Proposition 1.1 to prove that for a rational prime p $K_{\sigma(g^p)} = K_{\sigma(g)^p}$ for each $g \in G$. Raising the equation of Claim 1.2 to the $p - th$ power we obtain

$$(\sigma(g))^p = g^p + x^p + x_0 + p \cdot z_0$$

with $x_0 \in (RG, RG)$ and $z_0 \in RG$. Note that $x^p \in (RG, RG) + p \cdot RG$. Therefore we get the equation

$$(\sigma(g))^p = g^p + x_2 + p \cdot z_1 \text{ with } x_2 \in (RG, RG) \text{ and } z_1 \in RG.$$

Denote by C the conjugacy class of g^p. Now apply tr_C to the right hand side of the equation. This yields the value $1 + p \cdot tr_C(z_1)$ which is different from zero, if pR is different from R.

On the other hand the claim 1.2 shows that there is precisely one conjugacy class C_1 of G such that $tr_{C_1}(\sigma(g)^p)$ is not zero. For, if $tr_{C_1}(\sigma(g)^p) \neq 0$, then it follows from the claim 1.2 that $tr_{C_1}(g_0) \neq 0$, where g_0 is the unique element of G with $\sigma(g_0) = \sigma(g)^p$. Therefore C_1 is the conjugacy class of g_0.

Consequently we get that $K_{\sigma(g)^p} = K_{g^p}$. Of course $K_{g^p} = K_{\sigma(g^p)}$ and this establishes the Proposition. $\qquad\square$

1.3. Digressions. We look a little bit more carefully at the calculations in the proof of Proposition 1.1 in case there is no class sum correspondence between H and G is available. Assume that H is a group basis of RG. We define a map α from the set of conjugacy classes of p-elements of H into the power set of G as follows.

Let K be a conjugacy class of H, $h \in K$ and assume that h is a p–element of H. Assume further that $pR \neq R$. Since G is an R-basis, we can express K_h^H uniquely in the form $\Sigma a_i K_{g_i}$. Define now

$$\alpha(K) = \{C(g_i); a_i \text{is not in} pR\},$$

where $C(g_i)$ denotes the conjugacy class of G corresponding to the class sum K_{g_i}.

1.4. *The elements* g_i *in the definition of* $\alpha(K)$ *have the same order as* h.

PROOF: This is clear, if h is the trivial element of H. So we can use induction. Starting with $K(h) = \Sigma a_i K(g_i)$ we obtain $l_h \cdot h = \Sigma a_i \cdot l_i \cdot g_i + x$, where l_h and

l_i denote the length of the underlying conjugacy classes and $x \in (RG, RG)$. Consequently we obtain

$$l_h \cdot h^p = \Sigma \, a_i \cdot l_i \cdot g_i^p + y$$

with $y \in (RG, RG) + p \cdot RG$. Applying $tr_{C(g_i^p)}$ to this equation we see that

$$l_h \cdot tr_{C(g_i^p)}(h^p) = a_i \cdot l_i \cdot tr_{C(g_i^p)}(g_i^p) + p \cdot r$$

with $r \in R$.

1.5. Claim. $\alpha(K)$ *consists precisely of those classes* C *of* G *with* $tr_C(h) \cdot |K|/|C|$ *is not in* pR.

PROOF:

$$\sum_{x \in G} x^{-1} h x = \Sigma \, tr_{C(g_i)}(h) \cdot |C_G(g_i)| \cdot K_{g_i} = |C_H(h)| \cdot K_h.$$

Hence

$$K_h = \Sigma \, tr_{C(g_i)}(h) \cdot |C_G(g_i)| / |C_H(h)| \cdot K_{g_i}.$$

So the claim follows from the fact that the class sums of G form an R-basis of the centre $Z(RG)$. □

Now if a_i is not in pR, then $C_G(g_i^p)$ is in $\alpha(K_{h^p})$ by the claim. By induction we conclude that g_i^p has the same order as h^p. Hence g_i is a p-element of the same order as h. This completes the proof of 1.4. □

1.6. Proposition. *Let* H *be a group basis of* RG *and assume that there exists a class sum correspondence* $\sigma : G \to H$. *Let* π *be a surjective ring homomorphism from* RG *to* $(S)_n$, *the* $n \times n$-*matrix ring over an integral domain* S. *Then* $tr(\pi(g)) = tr(\pi(\sigma(g))$, *if the length of the conjugacy class of* g *is different from zero in* S.

PROOF: Since π is surjective, the centre of RG maps into the centre of $(S)_n$. Consequently $\pi(K(g)) = s \cdot I_n$. Since the trace of a matrix is invariant under conjugation, we get that $l_g \cdot tr(\pi(g)) = s \cdot n$ for some $s \in S$. Analogously it follows that $s \cdot n = l_{\sigma(g)} \cdot tr(\pi(\sigma(g)))$. Because $l_g = l_{\sigma(g)}$ and because S is an entire ring we can cancel, if l_g is different from 0. □

PROOF OF THEOREM V.1 (b) — (d): Since by section IV.1 part (a) of the Theorem holds we may apply Proposition 1.1 and get immediately (c).

From (c) it follows that g and $\sigma(g)$ have the same order. The augmentation map shows that $C(g)$ and $C(\sigma(g))$ have the same length. Hence (b) is established. Finally, if K is a field containing R, the Wedderburn decomposition of KG and Proposition 1.6 prove (d). □

§2   Character tables

Next we consider properties of a finite group G reflected by its ordinary character table $CT(G)$. In the sense of R.Brauer $CT(G)$ is an $h \times h$ — matrix $M = (m_{ij})$ of algebraic integers, where the columns are labelled by the conjugacy classes C_j of G, the rows are labelled by the ordinary irreducible characters χ_i, and $m_{ij} = \chi_i(C_j)$. It is a usual convention that the first row is labelled by the trivial character and the first column by the class of the trivial element of G. In contrast to common computer algebra systems we do not assume that the headline of the character table is given. In particular the order of the elements of a conjugacy class is unknown and also the power map on the classes. In general the character table does not determine the group up to isomorphism. The standard counterexample is given by the two nonisomorphic nonabelian groups of order p^3. Next we collect some elementary facts about properties reflected by $CT(G)$. For further results and a more detailed description see the book of Isaacs [Is; 76].

2.1. Theorem. *The character table $CT(G)$ of a finite group G determines*

a) the length of the conjugacy classes,

b) constructively the lattice of normal subgroups and the order of the normal subgroups. The normal subgroups which are kernels of irreducible characters are determined. Moreover the conjugacy classes whose unions form a normal subgroup are known.

c) For each normal subgroup N of G the character table $CT(G/N)$ may be constructively computed from $CT(G)$.

PROOF: a) follows immediately from the orthogonality relations.

b) A conjugacy class C_j lies in the kernel of χ_i if, and only if, $m_{ij} = m_{i1} = \deg(\chi_i)$. Therefore the kernels of the irreducible characters are known as unions of conjugacy classes and the lattice of the normal subgoups is obtained by calculating all the intersections of these kernels. Hence all normal subgroups are given as union of conjugacy classes and by a) their order is immediately determined.

c) Let N be the union of the classes $C_{\alpha_1}, ..., C_{\alpha_n}$. Then $CT(G/N)$ is obtained from $CT(G)$ by the following algorithm.

Delete all characters χ which do not have the same value on all the classes C_{α_i}. After this select one class under the classes which are equal on the remaining characters and remove the other ones.

The algorithm works since each irreducible character χ of G/N gives rise to one of G by composing χ with the reduction map $\pi : G \to G/N$. Vice versa, if N is in the kernel of χ, then χ arises in this way from G/N. Moreover the image of a conjugacy class under π is a conjugacy class of G/N.   □

The object of the second part of this section is the following result.

2.2. Theorem. *[Ki; 91] The character table of a finite group G determines the chief series of G.*

Theorem V.1 (f) is an immediate consequence of Theorem V.1 (d) and Theorem 2.2. The latter is proved with the following result which depends on the classification of the finite simple groups.

2.3. Theorem. *[K-L-S-T; 90] Let S and T be finite simple groups. Assume that $|S|^a = |T|^b$ for some $a, b \in \mathbb{N}$. Then $a = b$.*

Moreover it follows that $S \cong T$ except in the cases $\{S, T\} = \{A_8, A_2(4)\}$ or $\{S, T\} = \{B_n(q), C_n(q)\}$ for $n \geq 3$ and odd q. Here we used the Chevalley notation for the simple groups of Lie type.

For a complete proof of Theorems 2.3 and 2.2 we refer to [K-L-S-T; 90] and [Ki; 91]. Nevertheless we sketch the main arguments.

PROOF (SKETCH): By induction it suffices for Theorem 2.2 to show that the minimal normal subgroups of G are determined by $CT(G)$ up to isomorphism. Let M be such a minimal normal subgroup. Then $M \cong S^a$ for some $a \in \mathbb{N}$, where S is a simple group.

If S is of prime order, the result is clear. We divide the noncyclic simple groups into three parts: the alternating ones, the groups of Lie type and the sporadic ones. In order to establish the Theorem we have to play essentialy six games, alternating versus alternating, alternating versus sporadic, and so on.

The basic tool for the proof of Theorem 2.3 is of number theoretical nature. If X is a finite group and $p \in \pi(X)$, let $|X| = p^{c(X)}$ (so in general $c(X)$ is just a real number). Let $p^{d(X)}$ be the highest natural power of p dividing $|X|$. Define $\alpha_p(X) = c(X)/d(X)$ and $\alpha(X) = min_{p \in \pi(X)} \alpha_p(X)$.

Let T be an other simple group with $|T|^b = |S|^a$. In order to establish the first assertion $a = b$ of Theorem 2.3 note that $\alpha_q(S^a) = \alpha_q(S)$ and $\alpha_q(T^b) = \alpha_q(T)$ for each prime $q \in \pi(S)$. Clearly $|S|^a = |T|^b$ implies $\pi(S) = \pi(T)$ and thus $\alpha(S) = \alpha(T)$.

E.Artin proved 1955 for the simple groups known at that time that the functions α_p and α determine such a group up to isomorphism except in the cases stated in Theorem 2.3 [Ar; 84]. We remark that for the most cases it suffices to consider the function α. Note also that the exceptional cases were already known around 1900 [Di; 01].

In the final step of the proof of Theorem 2.2 the exceptional pairs have to be considered. So we may assume that the character table of G is given and we know already that G has a minimal normal subgroup M of order $|M| = |B_n(q)|^a$ or $|M| = |A_8|^a$. Moreover we know which of the conjugacy classes C of G labelling the rows of the character table $CT(G)$ of G belong to M. Let x be a representative of C. By a Theorem of G.Higman $CT(G)$ determines constructively the primes dividing the order of x [Is; 76, (8.21)]. In particular the conjugacy classes of 2-elements in M are determined.

Let $M \cong S^a$. Since the minimal normal subgroup M is perfect, the minimal normal subgroups of M are precisely the simple factors S of M. Write an element of M as an a–tuple $(x_1, ..., x_a)$.

Now one sees easily that such G-conjugacy classes of 2-elements in M have the smallest length, if they have a representative of type $(y, 1, ..., 1)$, where y is an involution of S with conjugacy classes of smallest length.

The following facts now complete the proof of Theorem 2.2.

a) Assume that $|M| = |A_8|^a$. Then $M \cong A_8^a$ if, and only if, there exists a G-conjugacy class in M of length $a \cdot 105$. Otherwise $M \cong PSL(3,4)^a$.

b) Assume that $|M| = |B_n(q)^a|$. Then $M \cong B_n(q)^a$ if, and only if, there exists a G-conjugacy class in M of length

$a \cdot q^n \cdot (q^n + \epsilon)/2$ with $q^n \equiv \epsilon \mod 4$ and $\epsilon \in \{1, -1\}$.

Otherwise $M \cong C_n(q)^a$.

$\square$

§3   Unit groups of rings

The integral group ring $\mathbb{Z}G$ has the following universal property.

Let S be a ring and let U a subgroup of S^*. Assume that there exists a group homomorphism $\tau : G \to U$. Then τ lifts uniquely to a ring homomorphism τ_1 from $\mathbb{Z}G$ to S.

The following proposition may be derived from [San; 74]. For a more detailed discussion in particular with respect to circle groups see [San; 74].

3.1. Proposition. *Assume that for a given group G there exists a ring S such that all subgroups of S^* of order $|G|$ are isomorphic to G and such that G is in fact isomorphic to a subgroup of S^*, then the isomorphism problem of $\mathbb{Z}G$ has a positive solution.*

PROOF: By the universal property of $\mathbb{Z}G$ there exists a ring homomorphism $\tau : \mathbb{Z}G \to S$ such that $\tau(G) \cong G$. Let H be a group basis of $\mathbb{Z}G$. We show that $\tau(H)$ has order $|G|$. Then the assumption yields that $H \cong G$.

Suppose that $\tau_{|H}$ is not injective. Then this map has a non-trivial kernel M. By the normal subgroup correspondence we find a normal subgroup N of G with $I(N) \cdot \mathbb{Z}G = I(M) \cdot \mathbb{Z}G$. Clearly $\tau(I(M) \cdot \mathbb{Z}G) = 0$. Hence $\tau(I(N) \cdot \mathbb{Z}G) = 0$. Whence N is contained in the kernel of $\tau_{|G}$ contradicting that $\tau_{|G}$ is an isomorphism. $\square$

As a first application of Proposition 3.1 we prove Whitcomb's result that the isomorphism problem for $\mathbb{Z}G$ has a positive solution, if G is metabelian [Wh; 68]. For a different proof we refer to section XI.

3.2. Theorem. *Assume that G is metabelian. Then $\mathbb{Z}G \cong \mathbb{Z}H$ implies $G \cong H$.*

PROOF: By assumption, G has a normal subgroup A such that $B = G/A$ is abelian. This gives rise to the exact sequence

$$1 \to \mathbb{Z}G \cdot I(A) \to \mathbb{Z}G \to \mathbb{Z}B \to 1.$$

Factoring out $I(G) \cdot I(A)$ we get

$$1 \to \mathbb{Z}G \cdot I(A)/I(G) \cdot I(A) \to \mathbb{Z}G/I(G) \cdot I(A) \to \mathbb{Z}B \to 1.$$

We remark that the ring $\mathbb{Z}G/I(G) \cdot I(A)$ is often called the small group ring with respect to A denoted by $S(G, A)$. We show that $S(G, A)$ satisfies the assumptions of Proposition 3.1

$\square$

3.3. Lemma. *Let G be a not necessarily finite group.*
a) If $H \leq G$, then as abelian groups $\mathbb{Z}G \cdot I(H)/I(G) \cdot I(H) \cong H/[H, H]$.
b) If A is an abelian normal subgroup of G, then A may be regarded as right $\mathbb{Z}G$ or $\mathbb{Z}G/A$-module via conjugation (i.e. the action is given by $g * a = g^{-1}ag$). Moreover A is isomorphic to $\mathbb{Z}G \cdot I(A)/I(G) \cdot I(A)$ (regarded as $\mathbb{Z}G$ – module by right multiplication) as $\mathbb{Z}G$-module, $\mathbb{Z}G/A$-module rsp.

PROOF: If T is a transversal of H in G, then $\mathbb{Z}G \cdot I(H)$ has a $\mathbb{Z}$-basis of the form $\{g_i(h - 1); g_i \in T, h \in H^{\#}\}$. Sending $g_i(h - 1)$ to $h \cdot [H, H]$ defines a surjective $\mathbb{Z}$-homomorphism μ from $\mathbb{Z}G \cdot I(H)$ to $H/[H, H]$. Look at the following calculation.

$$(g - 1)(h - 1) = g(h - 1) - (h - 1)$$

As $I(G) \cdot I(H)$ is generated by $\{(g - 1)(h - 1); h \in H, g \in G\}$, it follows that $I(G) \cdot I(H)$ is contained in the kernel of μ. So μ factors through the quotient $\mathbb{Z}G \cdot I(H)/I(G) \cdot I(H)$.
The map
$\sigma : H/[H, H] \to \mathbb{Z}G \cdot I(H)/I(G) \cdot I(H)$ given by $h \cdot [H, H] \to h - 1 + I(G) \cdot I(H)$
is well-defined. To see this consider $hkh^{-1}k^{-1} + I(G) \cdot I(H)$. Note that modulo $I(G) \cdot I(H)$ the following rule holds

$$(ab - 1) = -(a - 1) - (b - 1).$$

Hence, for $h, k \in H$, modulo $I(G) \cdot I(H)$

$$hkh^{-1}k^{-1} - 1 = -(h - 1) - (k - 1) - (h^{-1} - 1) - (k^{-1} - 1) = 0.$$

One checks easily that σ is a homomorphism and that $\mu\sigma = id$.
b) If H is normal, then $I(G) \cdot I(H)$ and $\mathbb{Z}G \cdot I(H)$ are 2-sided ideals. If H is additionally supposed to be abelian, then the homomorphism constructed in a) is an isomorphism. Clearly $(h - 1) \cdot g = g(g^{-1}hg - 1)$ and this shows immediately that the isomorphism of a) is in fact a module isomorphism.

$\square$

PROOF OF THEOREM 3.2 CONTINUED:

As A is abelian it follows by Lemma 3.3 that

$$g \to g - 1 + I(G) \cdot I(A)$$

yields an embedding of G into $S(G, A)^*$. The Theorem follows from Proposition 3.1, if we have shown that there is no other subgroup of $S(G, A)^*$ with order $|G|$.

Note for this that the ideal $J = I(G)/I(G) \cdot I(A)$ is nilpotent. Therefore the projection from $S(G, A)$ onto $\mathbb{Z}B$ is surjective with respect to units, $i.e$ we have an exact sequence

$$1 \to 1 + J \to S(G, A)^* \to (\mathbb{Z}B)^* \to 1.$$

By Lemma 3.3 we obtain that $1 + J \cong A$. Since B is abelian, the class sum correspondence yields that $\mathbb{Z}B$ has precisely one group basis. Thus $S(G, A)^*$ has precisely one subgroup of order $|G|$.

$\square$

3.4. Remarks on infinite groups.

a) In Lemma 3.3 we can consider the special case that $H = G$ and that G is abelian. Then the first part of the Lemma yields the well known positive solution for all abelian groups via the isomorphism $I(G)/I(G)^2 \cong G/[G, G]$.

b) In contrast to finite abelian groups the class sum correspondence (or one could also say the Zassenhaus conjecture, c.f. section IX) does not hold in general for infinite abelian groups. A counterexample is $G = C_\infty \times C_5$, see [Wallace; 87].

However it holds for infinite abelian torsion groups. For this let G be such a group and let $u \in V(\mathbb{Z}G)$ be a unit of finite order. If $u = \Sigma_{g \in G} r_g \cdot g$, then only finitely many coefficients r_g are non-zero. Hence u is contained in the group ring of a finitely generated subgroup, i.e. in this case in a finite abelian subgroup of G. Hence $u \in G$ by G.Higman [Hig; 39].

If G is a torsionfree group, it is conjectured that all normalized units of $\mathbb{Z}G$ are group elements [Se; 78, p.164] or [De; 77, pp.112-113]. In particular if G is abelian torsionfree the conjecture holds. For further results we refer to [Se; 78] or [Pa; 77].

Next we discuss briefly the so-called splitting problem, i.e. the question whether there is a group homomorphism $\mu : V(\mathbb{Z}G) \to G$ such that $\iota\mu = id_G$, where ι denotes the inclusion from G into $V(\mathbb{Z}G)$.

3.5. Proposition.

 a) If G is the unit group of a ring, then the splitting-problem has a positive solution. Moreover the isomorphism problem has a positive solution for $\mathbb{Z}G$. In particular the groups $GL(n, q)$, q denotes as usual the order of the underlying finite field, have these properties.

b) The isomorphism problem has a positive solution for $G = SL(n, q)$.

PROOF:

a) This follows immediately from 3.1.

b) $SL(n, q)$ is a perfect group except that $n = 2$, $q = 2$ or $n = 2$, $q = 3$. The exceptions are handled by direct inspection. By Lemma 3.3 it follows that, if G is perfect, then any group basis of $\mathbb{Z}G$ is perfect. By the universal property of $\mathbb{Z}G$ we have a map

$\sigma : \mathbb{Z}SL(n, q) \to Mat(n, q)$ sending $SL(n, q)$ isomorphically to $SL(n, q)$.

By the normal subgroup correspondence the image of each group basis of $\mathbb{Z}SL(n, q)$ under σ is a perfect group of order $|SL(n, q)|$. Since $SL(n, q)$ is the maximal subgroup of $GL(n, q)$ being perfect, the result follows.

$\square$

3.6. Remarks.

a) In [R-S; 83] it is shown that the answer to the splitting problem is negative. The Frobenius group $G = C_{73} \rtimes C_8$ is one of their counterexamples, another one is $C_{241} \rtimes C_{10}$. Nevertheless for some metabelian groups the answer is positive. The most far reaching result for metabelian groups is the following of Cliff–Sehgal–Weiss [C-S-W; 81].

If G has an abelian normal subgroup of N with G/N abelian of odd order, then there is a splitting. Moreover the kernel of the splitting is torsionfree.

Clearly the counterexamples mentioned above indicate that the result of Cliff–Sehgal–Weiss is for metabelian groups in some sense best possible.

An essential ingredient for the counterexample with the group $C_{241} \rtimes C_{10}$ is that $\mathbb{Z}C_{10}$ has the following unit

$$
\begin{aligned}
u \;=\; & -372099 \cdot c^0 + 114985 \cdot c^1 + 301035 \cdot c^2 - 301035 \cdot c^3 - 114985 \cdot c^4 \\
& +372100 \cdot c^5 - 114985 \cdot c^6 - 301035 \cdot c^7 + 301035 \cdot c^8 + 114985 \cdot c^9.
\end{aligned}
$$

The unit u has the property that it maps to the identity under the map

$$ V(\mathbb{Z}C_{10}) \to V(\mathbb{Z}C_2) \times V(\mathbb{Z}C_2). $$

b) The situation is even worse in the case of simple groups.

In [R-S; 85] it is suspected that the following is true.

If G is a non-abelian simple group and if G is not isomorphic to $PSL(n, q)$ with $n \geq 2$, then the answer to the splitting problem is negative.

In fact the authors give evidence (including [R-S; 85, footnote p.398]) for this suspicion and show for many non-abelian simple groups G (e.g. A_n for $n > 8$) that $V(\mathbb{Z}G)$ does not allow a splitting.

One of the basic tools of the results in [R-S; 85] and [R-S; 83] is the use of the Congruence Subgroup Theorem of Bass–Milnor–Serre [B-M-S; 67]. The basic point is that, if there is a splitting for $V(\mathbb{Z}G)$, then this Theorem implies the existence of a normal subgroup N of $V(\mathbb{Z}G)$ such that $V(\mathbb{Z}G)/N$, up to a few modifications, may be considered as a subgroup of the units of a finite ring. The finite simple rings are known by Wedderburn's Theorem (plus the fact that finite skew fields are fields) and the non-abelian composition factors of unit groups of finite rings are $PSL(n, q)'s$.

c) With respect to nilpotent groups the splitting problem seems to be open. Note the following.

If $\mu : V(\mathbb{Z}G) \to G$ is a splitting and G is nilpotent, then it follows that the isomorphism problem for G has a positive solution.

For let H be a group basis of $\mathbb{Z}G$. Assume that μ restricted to H is not injective. Then, as H is nilpotent, μ has to map a non-trivial element of the centre $Z(H)$ to 1. However $Z(G) = Z(H)$ by the class sum correspondence. Hence μ restricted to H has to be injective and, as $|H| = |G|$, surjective.

d) Note that the splitting problem for $V(\mathbb{Z}G)$ has a positive solution if, and only if, there exists a ring R such that G is isomorphic to a subgroup U of R^* and U has a normal complement in R^*.

e) **Crystallographic groups.** Another interesting topic in the context of unit groups of rings and the isomorphism problem of integral group rings are the finite subgroups of $GL(n, \mathbb{Q})$ or $GL(n, \mathbb{Z})$, in particular the maximal finite subgroups. For recent progress on the determination of such groups we refer to [Pl; 91]. Note that Proposition 3.1 implies immediately that, if such subgroups G are up to isomorphism determined by their order, then the isomorphism problem for $\mathbb{Z}G$ has a positive solution.

VI Subgroup rigidity

As we will see in Section IX, the isomorphism problem is closely related to the study of augmentation preserving automorphisms of the group ring (cf. the discussion of the Zassenhaus conjecture in [R-S; 87 3]), and to the structure of the group of augmented units in the group ring. We shall quote here without proof the most far reaching result, which has been obtained so far (cf. [Sc; 87] and [Sc; 90]).

VI.1. Theorem. *Let G be a finite group and $p > 0$ a rational prime. Assume that G has a normal p-subgroup N with $C_G(N) \leq N$. If $\alpha : \mathbb{Z}G \longrightarrow \mathbb{Z}G$ is an augmented automorphism[14], then there exists a unit $u \in V(\hat{\mathbb{Z}}_p G)$ such that $\alpha(G) = u \cdot G \cdot u^{-1}$. In particular, the Zassenhaus conjecture is true for the above class of groups[15].*

VI.2. Remarks.

- In case G is solvable, then the above condition is equivalent to the condition, that $O_{p'}(G) = 1$; i.e. G has no normal subgroup of order prime to p. This in turn is equivalent that $\hat{\mathbb{Z}}_p G$ consists of the principal block only (cf. section II Lemma 1.1).
- The condition that α is an automorphism of $\mathbb{Z}G$ is actually much too strong. It is needed to ensure, that the induced automorphism

$$\alpha_p : \hat{\mathbb{Z}}_p G \longrightarrow \hat{\mathbb{Z}}_p G$$

preserves the induced augmentation ideal $I_{\mathbb{Z}}(N)G$ of N in $\hat{\mathbb{Z}}_p G$, and can be replaced by this latter condition. (I do not know whether this condition is automatic.)
- The proof involves Green correspondence of automorphisms and one of the main ingredients is the "subgroup rigidity theorem", of which an extended version with proof is presented below.

In this connection I would also like to mention the following result of L. L. Scott [Sc; 90].

VI.3. Theorem. *Let R be a Dedekind domain of finite rank over $\hat{\mathbb{Z}}_p$. Let B be the principal block of the two group algebras RG and RH over R, for finite groups G and H. Let D and E be Sylow p-subgroups of G and H, resp., and identify them with their projections into B. Assume that E is augmented (the augmentation being induced from that of RG). Assume that D is cyclic (which implies that E is cyclic), and that D and E are T.I. sets in G and H, resp. Then D is conjugate to E in B.*

---

14   i.e. $\epsilon_G \alpha = \epsilon_G$, that is α commutes with the augmentation map.
15   i.e. every augmented automorphism of $\mathbb{Z}G$ can be written as a group automorphism followed by a central automorphism. For details we refer to section IX.

We now turn to the *subgroup rigidity result for p-groups*:

§1   Statement of the results

Let R be a complete Dedekind domain of characteristic zero with residue field of characteristic $p > 0$, and let G be a finite p–group. It was shown in [R-S; 87 1], that given a finite subgroup U of the same order as G in the normalized units $V(RG)$ – i.e. units of augmentation one – of the group ring RG of G over R, then U is conjugate in $V(RG)$ to G. In that paper a considerable amount of work had to be done to pass from an unramified extension of $\hat{Z}_p$, the p–adic integers, to a ring as general as R above. It was proved in [We; 87] – a sketch can also be found in [Sc; 87] [16] – that for R an unramified extension of $\hat{Z}_p$, every finite subgroup U of RG is conjugate in $V(RG)$ to a subgroup of G. We shall prove here an extended version of this result to general rings R as above. Also here the generalization is non-trivial and yields an interesting generalization of a result of Weiss [We; 87]. The same general result was obtained by G. Thompson in his thesis [Th; 89].

Let me briefly point out, where the difficulty with ramification lies: Let C be a cyclic group of order p. If R is an unramified extension of $\hat{Z}_p$, let ζ be a primitive $p - th$ root of unity. Then the group ring RC is the pullback

$$
\begin{array}{ccc}
RC & \longrightarrow & R \\
\downarrow & & \downarrow \\
\Lambda & \longrightarrow & R/(p \cdot R)
\end{array}
\quad,
$$

where $\Lambda = R[\zeta]$ is a finite extension of R; moreover, the only indecomposable RC–lattices are R, Λ, RC. So one can prove statements about lattices by inspection, as Weiss has done in [We; 87].

If R now has ramification, then RC is still a pullback as above, however, Λ is not an order in a field any more: it is a rather complicated order, and in general, there are infinitely many indecomposable RC-lattices, RC is wild in most of the cases.

We give in Theorem 1.3 necessary and sufficient conditions, that a monomial representation of $R/rad^t(R)$ is the reduction mod $rad^t(R)$ of a monomial RG–lattice in case the field of fractions of R is a splitting field for G and all of its subgroups. This is used in Theorem 1.2 to give a criterion for when an RG–lattice M is a permutation module in terms of the restriction to a normal subgroup N and the structure of the module of N–fixed points M^N. The subgroup conjugacy is then an immediate consequence of Theorem 1.2.

Parts of the proof of Theorem 1.3 arose in a discussion with Wolfgang Kimmerle.

We shall use the following *notation*:

---

16   In his Ph. D. Thesis at the University of Virginia, 1989, Gary Thompson worked out the details of [Sc; 87], and extended them to the general Dedekind domain R above.

- R is a complete Dedekind domain of finite rank over $\hat{\mathbb{Z}}_p$, the p–adic integers,
- K is the field of fractions of R ,
- π is a parameter of R ; $i.e.$ $rad(R) = \pi \cdot R$,
- $R^\times$ is the multiplicative group of units of R ,
- R/π^t is the reduction of R modulo π^t for some fixed $t \in \mathbb{N}$,
- M/π^t is the reduction of an R–module M modulo π^t ,
- G is a finite p–group,
- χ_i^U are the various linear characters from the subgroup U of G to $R^\times$; $i.e.$ homomorphisms $U \longrightarrow R^\times$, $1 \le i \le n(U)$,
- χ_i^U/π^t are the various characters from U to $(R/\pi^t)^\times$, which are induced from the χ_i^U (note that different χ_i^U's may reduce to the same χ_j^U/π^t),
- $I_i^U(R)$ is the kernel of the map from RU to R induced by χ_i^U. This kernel is generated over R by $\{u - \chi_i^U(u)\}_{u \in U \setminus 1}$,
- $R_{\chi_i^U} \uparrow_U^G$ is the corresponding transitive generalized permutation module for G.

The notation of transitive generalized permutation module for $R/\pi^t G$ implies by the above, that the U–action on R/π^t is induced from an action on R . A *generalized permutation module* is a direct sum of transitive generalized permutation modules, for various subgroups. A similar notation is used for other rings of coefficients.

The following is an extension of Weiss' result [We; 87, Theorem 3], of which we present a proof, which is shorter than Weiss' – even in the unramified situation, provided one takes in that case Weiss Theorem 2 [We; 87] for granted.

1.1. Theorem (Subgroup rigidity). *Let $V(RG)$ be the units in RG of augmentation one, and G a p–group. If V is a finite subgroup of $V(RG)$, then*

$$u \cdot V \cdot u^{-1} \subset G \text{ for some unit } u \text{ in } RG .$$

The above result will be an immediate consequence of

1.2. Theorem. *Let M be an RG-lattice, and let N be a normal subgroup of G. Assume that*

  a.) $M \downarrow_N$ is a free RN–module,

  b.) $M/I_R(N) \cdot M$ is a permutation module for G/N .

Then M is an RG–permutation module.

Theorem 1.3 plays a major role in the proof of Theorem 1.2, though it is also of interest for its own sake. The notation is the one introduced above.

1.3. Theorem. *1.) Assume that K is a splitting field for G and all of its subgroups. Let ζ be a primitive p^{th} root of unity in R , then we require $\pi^t \cdot R \subset (1 - \zeta) \cdot R$.*

An RG−lattice M is a generalized permutation module if and only if M/π^t is a generalized permutation module – note that we are always assuming, that the various U−actions on R/π^t are induced from the U−actions on R.
2.) If π^t · R ⊄ (1 − ζ) · R , then there is an RG−lattice M, which is not a generalized permutation module, but M/π^t is a generalized permutation module.

1.4. Remarks to Theorem 1.3.

- If $R = \hat{\mathbb{Z}}_p[\zeta]$, where ζ is a primitive p-th root of unity and if $t = 1$, then the above conclusion is Weiss' Theorem 3 [We; 87].

- The above Theorem 1.3 is false, if K is not a splitting field for G and its subgroups. Let $n \geq 2$ and p an odd prime. Let ξ be a primitive $p^n - th$ root of unity and set $K = \hat{\mathbb{Q}}_p(\xi + \xi^{-1})$. Letting R be the ring of integers of K and $rad(R) = \pi \cdot R$, we take G cyclic of order p . Let S be the ring of integers in $K(\zeta) = \hat{\mathbb{Q}}_p(\xi)$ and set $q = rad(S)$. Letting G act on S by multiplication with $\zeta = \xi^{p^{n-1}}$, then S is an irreducible RG−lattice of rank 2 , such that G acts trivially on $S/\pi S$, since

$$(1 - \zeta) \cdot S = q^{p^{n-1}} \subset q^2 = \pi \cdot S;$$

thus $S/\pi \cdot S$ is a permutation module, but S is not a generalized permutation lattice. [17]

1.5. Note.

- The above theorem 1.1 states that the full Sylow theorems for finite p-subgroups hold in the profinite p-group $V(RG)$; a surprising property. It would be interesting to have a group theoretical criterion on a profinite p-group $\hat{Q}$, such that the Sylow theorems hold for the finite p-subgroups in $\hat{Q}$.

- In an attempt to prove 1.1 the next result, which gives a connection between the Sylow theorems and the cohomology variety of $V(RG)$ was obtained.

1.6. Theorem. *The following conditions are equivalent for a finite p-group G.*

- *Every finite p-subgroup U of V(RG) is conjugate in V(RG) to a subgroup of G.*

- *For every subgroup P of G, the natural inclusion*

$$N_G(P)/P \longrightarrow N_{V(RG)}(P)/P$$

induces a continuous map

$$V(N_G(P)/P, I\!\!F) \longrightarrow V(N_{V(RG)}(P)/P, I\!\!F),$$

which is a bijection.

---

17   This example was communicated to me by the referee of [Ro; 92].

- *The variety of $N_{V(RG)}(P)/P$ is connected for every subgroup P of G.*

(This result uses heavily Quillen's description of the cohomology variety [Qui; 71])

We shall derive now 1.1 from 1.2, since the argument is very short - this is the same argument as in [We; 87]:

PROOF OF THEOREM 1.1 FROM THEOREM 1.2: Let V be a finite subgroup in $V(RG)$. We let $\hat{G} = V \times G$, and consider $M = RG$ as a $R\hat{G}$–module via the action

$$(v, g) \cdot m = v \cdot m \cdot g^{-1} .$$

Then Theorem 1.2 can be applied $(N = G)$; note the point, where we use that V is a group of normalized units: as V–module $M/I(G) \cdot M = R$ is trivial only if V is normalized. Thus M must be a transitive permutation module, $M \simeq R \uparrow_H^{\hat{G}}$, where $H \cap G = 1$, and $|H| = |V|$; thus $h = (v_h, g_h)$, $v_h \in V$, $g_h \in G$, and so, if m_0 corresponds in M to $1 \in RG$ i.e. the coset H in $R\uparrow_H^{\hat{G}}$. Then m_0 is a unit in $M = RG$, and $v_h \cdot m_0 \cdot g_h^{-1} = m_0$; i.e. $m_0^{-1} \cdot v_h \cdot m_0 \in G$ for every $v_h \in V$; but $\{v_h\}_{h \in H} = V$. Whence the statement. q.e.d.   □

§2   The Proof of the First Theorem

The Proof of Theorem 1.3 is done by an induction on $|G|$ and the R–rank of M as in [We; 87].

Before we come to the actual proof we need an observation, where the restrictive condition on $R/(\pi^t \cdot R)$ becomes apparent:

2.1. Lemma. *Let R be a complete Dedekind domain of finite rank over $\hat{\mathbb{Z}}_p$. The induced map*

$$\kappa_G : Ext^1_{RG}(M, N) \longrightarrow Ext^1_{R/\pi^t G}(M/\pi^t \cdot M, N/\pi^t \cdot N)$$

is injective for all generalized permutation lattices M and N, if and only if either R does not contain a primitive p^{th} root of unity or $\pi^t \cdot R \subset (1 - \zeta) \cdot R$, where ζ is a primitive p^{th} root of unity.

PROOF: We note that $Ext^1_{RG}(-, -)$ is additive in both variables, and reduction modulo π^t is exact, since we are working with lattices. Thus we may assume that $M = R_i \uparrow_U^G$ and $N = R_j \uparrow_V^G$, where R_i and R_j resp. are one-dimensional representations, induced from the linear characters χ_U^i of U and χ_V^j of V resp.; moreover induction commutes with taking duals, and so by Frobenius reciprocity [18] we have:

$$Ext^1_{RG}(R_i \uparrow^G_U \ , R_j \uparrow^G_V) \simeq Ext^1_{RU}(R_i, R_j \uparrow^G_V \downarrow_U) \ .$$

By Mackey's Theorem[19],

$$R_j \uparrow^G_V \downarrow_U \simeq \oplus_x R_j \uparrow^U_{V \cap x U x^{-1}} \ .$$

Now we look at the various summands and apply Frobenius reciprocity again; thus, it is enough to find a necessary and sufficient condition, such that

$$\kappa_U : Ext^1_{RU}(R_i, R_j) \longrightarrow Ext^1_{R/\pi^t U}(R_i/\pi^t \cdot R_i, R_j/\pi^t \cdot R_i)$$

is injective for every subgroup U of G , and all one-dimensional representations R_i and R_j , induced from the characters χ^U_i and χ^U_j of U . Since for every one-dimensional representation R_k , the tensor product $R_k \otimes_R -$ is exact on $R-$lattices, there is no loss of generality, if we assume that $R_i = R$ is the trivial $RU-$lattice. If also $R_j = R$, then $Ext^1_{RU}(R, R) = 0$, and the statement follows. This also treats the case, where R does not contain a primitive $p - th$ root of unity. So we assume from now on, that $R_j \neq R$, and that R_j is given by the character

$$\chi^U_j : U \longrightarrow R^\times \ ,$$

which has kernel N , and $U/N = < \hat{u}_0 : \hat{u}_0^{p^m} >$ is cyclic of order p^m ; moreover, χ^U_j is given by sending u_0 , an inverse image of $\hat{u}_0$ in U , to ξ , a fixed primitive $p^m - th$ root of unity in R . By ξ/π^t we denote the image of ξ in $R/\pi^t \cdot R$. Since N lies in the kernel of any exact sequence in $Ext^1_{RU}(R, R_j)-$ note that this is just a matter of $K \otimes_R R$ and $K \otimes_R R_j-$ we may assume that $N = 1$ and U is cyclic of order p^m. We then have the commutative diagram with exact rows

$$
\begin{array}{ccccccccc}
0 & \longrightarrow & I^U(R) & \longrightarrow & RU & \longrightarrow & R & \longrightarrow & 0 \\
 & & \downarrow & & \downarrow & & \downarrow & & \\
0 & \longrightarrow & I^U(\overline{R}) & \longrightarrow & \overline{RU} & \longrightarrow & \overline{R} & \longrightarrow & 0
\end{array} \ ,
$$

where the vertical maps are induced from reduction modulo $\pi^t \cdot R$, and we have written $\overline{R}$ for $R/\pi^t \cdot R$. From the above diagram we obtain an induced commutative diagram with exact rows:

$$
\begin{array}{ccccccccccc}
0 & \rightarrow & (R, R_j)^0 & \rightarrow & (RU, R_j)^0 & \overset{\delta}{\rightarrow} & (I^U(R), R_j)^0 & \overset{\epsilon}{\rightarrow} & (R, R_j)^1 & \rightarrow & 0 \\
 & & \downarrow & & \downarrow \alpha & & \downarrow \beta & & \downarrow \gamma & & \\
0 & \rightarrow & (\overline{R}, \overline{R}_j)^0 & \overset{\lambda}{\rightarrow} & (\overline{RU}, R_j)^0 & \overset{\mu}{\rightarrow} & (I^U(\overline{R}), \overline{R}_j)^0 & \overset{\nu}{\rightarrow} & (\overline{R}, \overline{R}_j)^1 & \rightarrow & 0
\end{array}
$$

---

18   This states that for a subgroup $U \leq G$ and for every $RG-$module M and every $RU-$module N we have

$$Hom_{RG}(M, N \uparrow^G_U) \simeq Hom_{RU}(M \downarrow^G_U, N),$$

the above result follows by using derived functors.

19   This states that

$$M \uparrow^G_U \downarrow_L \simeq \oplus_{U \backslash G / L} (g \otimes M) \downarrow_{U \cap \ {}^g L} \uparrow^L \ .$$

where we have written $(-, -)^i$ for $Ext^i_{RU}(-, -)$ and $Ext^0_{RU}(-, -)$ is defined as $Hom_{RU}(-, -)$.

Now $(R, R_j)^0 = 0$, since $R \neq R_j$ and $(RU, R_j)^0 \simeq R_j$, $(I^U(R), R_j)^0 \simeq R_j$, U being cyclic and $R \neq R_j$. δ is multiplication with $(1 - \xi)$, and so

$$(R, R_j)^1 \simeq R/(1 - \xi) \cdot R .$$

Moreover,

$$((R/\pi^t \cdot RU), R_j/\pi^t)^0 \simeq R_j/\pi^t ,$$

$$(I^U(R/\pi^t \cdot R), R_j/\pi^t \cdot R_j)^0 \simeq R_j/\pi^t \cdot R_j ,$$

and so

$$(\overline{R}, \overline{R}_j)^1 \simeq (R/\pi^t \cdot R)/(((\xi - 1) \cdot R + \pi^t \cdot R)/\pi^t \cdot R) \simeq R/((\xi - 1) \cdot R + \pi^t \cdot R);$$

thus we have a commutative diagram with exact rows:

$$
\begin{array}{ccccccc}
R & \xrightarrow{\xi-1} & R & \longrightarrow & R/(1 - \xi) \cdot R & \simeq & (R, R_j)^1 \\
\downarrow \alpha & & \downarrow \beta & & \downarrow \gamma & & \downarrow \gamma \\
R/\pi^t \cdot R & \longrightarrow & R/\pi^t \cdot R & \longrightarrow & R/((\xi - 1) \cdot R + \pi^t \cdot R) & \simeq & (\overline{R}, \overline{R}_j)^1
\end{array}.
$$

The map

$$\gamma : R/(1 - \xi) \cdot R \longrightarrow R/((1 - \xi) \cdot R + \pi^t \cdot R)$$

is induced from the natural map $R \longrightarrow R/\pi^t \cdot R$. Now γ is injective if and only if

$$\pi^t \cdot R \subset (1 - \xi) \cdot R ,$$

in which case γ is an isomorphism. This must hold for every subgroup U of G. For U cyclic of order p, we have the condition

$$\pi^t \cdot R \subset (1 - \zeta) \cdot R = (1 - \xi)^{p^{m-1}} \cdot R \subset (1 - \zeta) \cdot R ,$$

where ζ is a primitive $p - th$ root of unity in R. Whence the statement of the lemma.

We note, that if $Ker(\gamma) \neq 0$, and if

$$0 \longrightarrow R_j \longrightarrow X \longrightarrow R \longrightarrow 0$$

is a non split sequence in $Ker(\gamma)$, then X is not a generalized permutation module, however $X/\pi^t \cdot X$ is a generalized permutation module, since it is the direct sum of $\overline{R}$ and $\overline{R}_j$.

This proves the lemma. q.e.d. □

We are now in the position to continue with the

PROOF OF THEOREM 1.3:

Note that now K, the field of fractions of R, is a *splitting field for G and all its subgroups*. The assumption is that $\pi^t \cdot R \subset (1 - \zeta) \cdot R$, and that M/π^t is a generalized permutation module.

We shall use induction on $|G|$ and on the R−rank of M.

So let M be an RG-lattice with $M/\pi^t \cdot M$ a generalized permutation module for $(R/\pi^t \cdot R)G$; i.e.

$$M/\pi^t = \oplus_{i,U}((R/\pi^t \cdot R)_{\chi_i^U}) \uparrow_U^G .$$

We have to consider two cases:

2.2. Case 1 . The vertices[20] of all indecomposable summands of $M/\pi^t \cdot M$ are G ; *i.e.* $M/\pi^t \cdot M$ is a direct sum of one-dimensional $(R/\pi^t \cdot R)G$−modules.

Reduction to the case where M is irreducible:

If M is reducible, *i.e.* M contains an R−pure submodule M_0 of smaller rank, then the result follows by induction on the rank of M and thanks to the lemma.

In fact, we have the commutative diagram with exact rows:

(31)
$$
\begin{array}{ccccccccc}
0 & \longrightarrow & M_0 & \longrightarrow & M & \longrightarrow & M/M_0 & \longrightarrow & 0 \\
 & & \downarrow & & \downarrow & & \downarrow & & \\
0 & \longrightarrow & \overline{M_0} & \longrightarrow & \overline{M} & \longrightarrow & \overline{M/M_0} & \longrightarrow & 0
\end{array}
$$
,

where the vertical maps are reduction modulo π^t , and " $^-$ " is reduction modulo π^t . Since M_0 is an R−pure submodule of M, the $R/\pi^t \cdot R$−modules $\overline{M_0}$ and $\overline{M/M_0}$ are $R/\pi^t \cdot R$−free, and so the bottom sequence is $R/\pi^t \cdot R$−split. Moreover, $\overline{M}$ is the direct sum of one dimensional $(R/\pi^t \cdot R)G$−modules. It remains to show that the bottom sequence is $(R/\pi^t \cdot R)G$−split. This is a general phenomenon:

2.3. Claim. *Let*

$$0 \longrightarrow X' \xrightarrow{\alpha} X \xrightarrow{\beta} X'' \longrightarrow 0$$

be an exact sequence of $(R/\pi^t \cdot R) \cdot G$−modules, which is $R/\pi^t \cdot R$−split and with X an $(R/\pi^t \cdot R) \cdot G$−direct sum of modules of the form $(R/\pi^t \cdot R) \cdot e_i$, which are $R/\pi^t \cdot R$−free of rank one. Then the above sequence is $(R/\pi^t \cdot R)G$−split exact.

PROOF OF CLAIM 2.3: Let $X = \oplus_{i=1}^n (R/\pi^t \cdot R) \cdot e_i$, and denote by $\overline{X}$ and $\overline{X''}$ reduction modulo $\pi \cdot (R/\pi^t \cdot R) = rad(R/\pi^t \cdot R)$. Then $\overline{X}$ and $\overline{X''}$ are trivial $(R/\pi^t \cdot R) \cdot G$−modules. The map β induces a map

$$\overline{\beta} : \overline{X} = \oplus_{i=1}^n (\overline{R/\pi^t \cdot R}) \cdot \overline{e_i} \longrightarrow \overline{X''} ,$$

---

20   The smallest group V – up to conjugation – such that $M \oplus X \simeq N \uparrow_V^G$ for some RV−lattice N and some X.

which is now just a map of vector spaces. Hence there exists a subset $\{e_{i_1}, ..., e_{i_\nu}\}$ of $\{e_i\}_{1 \leq i \leq n}$, such that

$$\overline{\beta} : \overline{X} = \oplus_{j=1}^{\nu}(\overline{R/\pi^t \cdot R}) \cdot \overline{e}_{i_j} \longrightarrow \overline{X''} ,$$

is an isomorphism. By Nakayama's lemma the $(R/\pi^t \cdot R) \cdot G$−linear map

$$\beta : X = \oplus_{j=1}^{\nu}(R/\pi^t \cdot R) \cdot e_{i_j} \longrightarrow X'' ,$$

is surjective and consequently an isomorphism, X'' being $R/\pi^t \cdot R$−free. Hence we have constructed the desired splitting. This proves the Claim 2.3. q.e.d. □

We now return to the above commutative diagram (31). By the claim, $\overline{M} \simeq \overline{M_0} \oplus \overline{M/M_0}$ as $(R/\pi^t \cdot R) \cdot G$−modules. The Krull – Schmidt theorem shows, that $\overline{M_0}$ and $\overline{M/M_0}$ are generalized permutation modules, and by induction M_0 and M/M_0 are generalized permutation lattices. The lemma allows us to conclude that $M \simeq M_0 \oplus M/M_0$ is a generalized permutation lattice.

Thus we may assume that M is irreducible.

We shall show that this implies that M has R−rank one. Let U be a normal subgroup of index p. Then $M \downarrow_U = \oplus R_i$ by induction on $|G|$.

2.4. Claim. $M \downarrow_U$ is homogeneous; i.e. $M \downarrow_U \simeq R_j^{(n)}$ for some fixed homomorphism $\chi_j^U : U \longrightarrow R^\times$, which has kernel N, and U/N is cyclic of order p^m.

PROOF: By Clifford's theorem[21], $K \otimes_R M \downarrow_U \simeq (\oplus_i {}^{g_i}K_j)^e$, where ${}^{g_i}K_j$ are different conjugate representations of K_j ; we have to show $e = p$. We have

$$M \downarrow_U \simeq \oplus_i(\oplus_{k=1}^e {}^{g_i}(R_j \cdot m_k))$$

In particular, $g_i \cdot m_k$ must generate ${}^{g_i}(R_j \cdot m_k)$, since otherwise $g_i \cdot m_k \in \pi \cdot R$, and thus $m_k = g_i^{-1} \cdot g_i \cdot m_k \in \pi \cdot R_j \cdot m_k$, a contradiction. But then M can be generated by e elements as RG−module. On the other hand, $M/\pi^t \cdot M$ is as $(R/\pi^t \cdot R)G$−module a direct sum of one-dimensional modules, and so the minimal number of generators of M is p. Thus $M \downarrow_U$ must be homogeneous.

This proves the Claim 2.4. □

Since $M \downarrow_U$ is homogeneous, M has kernel N, and so it is a G/N−module. Note that N is normal in G, $M \downarrow_U$ being homogeneous.

Thus we can assume that $N = 1$ and then have two possibilities for G :

Type I: G is cyclic of order p^{m+1}. Since R contains primitive $p^{m+1} - th$ roots of unity – K is a splitting field for G and all subgroups – , all irreducible RG−lattices have rank one, and are generalized permutation lattices, and the statement follows.

---

21   cf. section XIII

Type II: G is of exponent p^m, it has a cyclic normal subgroup U of order p^m, and $|G : U| = p$. By Ito's theorem on the character degrees[22] - K is a splitting field for G and all subgroups - , every irreducible representation of KG has either rank one, in which case there is nothing to show, or it has rank p. Let V be an irreducible KG−module of K−dimension p. Then V occurs with multiplicity p in KG. However, $KG \downarrow_U = KU^{(p)}$, and every irreducible KU−module occurs with multiplicity one in KU. Thus $V \downarrow_U$ is multiplicity free. But in Claim 2.4 we had shown that $M \downarrow_U$ must be homogeneous, thus M must have R−rank one.

This finishes the argument of case 1.

2.5. Case 2. We assume now that

$$(32) \qquad M/\pi^t \cdot M = \oplus_{i,U}((R/\pi^t \cdot R)_{\chi_i^U}/\pi^t) \uparrow_U^G \ ,$$

here at least one $U \neq G$.

The arguments here are slight modifications of those of Weiss [We; 87]. For the convenience of the reader we give the details. We have to recall some results from Weiss' paper in a slightly modified version [We; 87]:

2.6. Definition. *[We; 87]* Let S be either R or a quotient $R/\pi^\mu \cdot R$ and assume that N is a generalized permutation module for SG. An SG split monomorphism

$$\vartheta : E \longrightarrow N$$

is called an *eigenfactor*, provided

  (a) the indecomposable factors of E have vertex G,

  (b) the indecomposable factors of $coker(\vartheta)$ have vertex a proper subgroup of G .

If N is a generalized permutation lattice for G, then an eigenfactor surely reduces modulo π^μ to an eigenfactor of $N/\pi^\mu \cdot N$.

2.7. Remark. Let $I\!\!F$ be a finite field of characteristic p , and let V be an $I\!\!FG$−module. For a subgroup H of G, we choose coset representatives $G = \cup g_i \cdot H$ and denote by

$$Tr_{G/H} : H^0(H, V) \longrightarrow H^0(G, V),$$

$$v \longrightarrow \Sigma \, g_i \cdot v$$

the relative trace map. We shall write V^H for $H^0(H, V)$ and note, that the trace is *associative*; i.e. if $U \leq H \leq G$, then $Tr_{G/U} = Tr_{G/H} \cdot Tr_{H/U}$.

---

22   This states, that if A is an abelian normal subgroup of G, then $\chi(1)|\,|G : A|$ for every irreducible character χ.

Moreover, the following interpretation is often useful: The unique natural $I\!\!FG$–homomorphism $\kappa_H : I\!\!F \longrightarrow I\!\!F \uparrow_H^G$, which maps $I\!\!F$ onto the socle of $I\!\!F \uparrow_H^G$ induces for every $I\!\!FG$–module W a homomorphism

$$\kappa_H^* : Hom_{I\!\!FG}(I\!\!F \uparrow_H^G, W) \longrightarrow Hom_{I\!\!FG}(I\!\!F, W):$$

By Frobenius reciprocity,

$$Hom_{I\!\!FG}(I\!\!F \uparrow_H^G, W) \simeq Hom_{I\!\!FH}(I\!\!F, W \downarrow_H) = H^0(H, W \downarrow_H)$$

and

$$Hom_{I\!\!FH}(I\!\!F, W) = H^0(H, W).$$

With this identification,

$$\kappa_H^* = Tr_{G/H}$$

2.8. Lemma. *[We; 87] Let V be an $I\!\!FG$–permutation module and let H be a subgroup of G .*

(a) *$Tr_{G/H}(V^H) \neq 0$ if and only if V has a direct summand isomorphic to $I\!\!F \uparrow_K^G$ for some subgroup K – up to conjugacy – of H*

(b) *Assume that H is minimal with respect to $Tr_{G/H}(V^H) \neq 0$. Let $v \in V^H$ with $Tr_{G/H}(v) \neq 0$.*

Then

$$\varphi_v : I\!\!F \uparrow_H^G \longrightarrow V,$$

$$g \otimes 1 \longrightarrow g \cdot v$$

defines a split monomorphism. The inverse map is given as follows: Let $V = \oplus_r I\!\!F \uparrow_{G_r}^G$ be a fixed decomposition into transitive permutation modules and write $v = \Sigma \, v_r$ according to this decomposition. Then $v_r \in V^H$ and there is an index r_0 such that $Tr_{G/H}(v_{r_0}) \neq 0$. The projection

$$\psi : V \longrightarrow I\!\!F \uparrow_H^G,$$

$$(x_r) \longrightarrow x_{r_0}$$

has the property, that $\psi \cdot \varphi_v$ is an isomorphism of $I\!\!F \uparrow_H^G$.

PROOF OF LEMMA 2.8:
(a): Because of Remark 2.7 $Tr_{G/H}(V^H) \neq 0$ iff $I\!\!F \uparrow_H^G$ injects into $V = \oplus_i I\!\!F \uparrow_{U_i}^G$. But this is only possible, if U_{i_0} is conjugate to a subgroup of H for some i_0 and conversely.
(b) is an immediate consequence of (a) and 2.7.   □

2.9. Lemma. *[We; 87] Let V be an $I\!\!FG$–permutation module, and let H be a subgroup of G, which is minimal with respect to $Tr_{G/H}(V^H) \neq 0$. If*

$$\vartheta : E \to V \downarrow_H$$

is an eigenfactor of $V \downarrow_H$ as $I\!FH-module$, then

$$Tr_{G/H}(V^H) = Tr_{G/H}(\vartheta(E)) \ .$$

PROOF OF LEMMA 2.9: Let $V = \oplus_i I\!F \uparrow_{U_i}^G$. Because of the minimality of H, no U_i is conjugately contained in H. Moreover,

$$V \downarrow_H = \oplus_{i,j} I\!F \ {}^{g_j}U_i \cap H,$$

when $G = \bigcup_i g_i H$. The eigenfactor E consists of those summands, where ${}^{g_j}U_i = H$ and $\vartheta(E) \subseteq V^H$. Because of Lemma 2.8 and because of the minimality of H, only $\vartheta(E)$ can give a contribution to $Tr_{G/H}(V^H)$. $\qquad\square$

We now *turn to the treatment of Case 2.* Let

$$M/\pi^t \cdot M = \oplus_{i,U}((R/\pi^t \cdot R)_{\chi_i^U}/\pi^t) \uparrow_U^G \ .$$

Choose $H < G$ minimal, such that $M/\pi^t \cdot M$ has a direct summand with vertex H. We denote by " $^-$ " the reduction modulo π. By induction, $M \downarrow_H$ is a generalized permutation lattice, and by the choice of H, $M \downarrow_H$ has a non trivial eigenfactor $0 \neq E = \oplus_i E_i e_i$, with $E_i \simeq R_{\chi_i^H}$,

$$(33) \qquad\qquad\qquad \vartheta : E \longrightarrow M \downarrow_H \ ;$$

then $E/\pi^t \cdot E$ and $\overline{E}$ are also eigenfactors for $M/\pi^t \cdot M$ and for $\overline{M}$ resp. (cf. 2.6). By 2.9 we have

$$Tr_{G/H}(\overline{\vartheta}(\overline{E})) = Tr_{G/H}(\overline{M}^H).$$

However, by construction of H, $Tr_{G/H}(\overline{M}^H) \neq 0$, and so $\overline{E}$ has an "eigenvector" $\overline{e} \in \overline{E}$ with $Tr_{G/H}(\overline{e}) \neq 0$. Since $\overline{e}_i$ is an $\overline{R}$−basis for $\overline{E}$, there exists an index i_0, such that $Tr_{G/H}(\overline{e}_{i_0}) \neq 0$.

2.10. Claim. $(R/\pi^t \cdot R)G \cdot \tilde{e}_{i_0}$ *is a direct summand of $M/\pi^t \cdot M$, isomorphic to* $((R/\pi^t \cdot R)_{\chi_{i_0}^H}) \uparrow_H^G$.

PROOF: Let $P = ((R/\pi^t \cdot R)_{\chi_{i_0}^H}) \uparrow_H^G$, and define

$$\alpha : P \longrightarrow M/\pi^t \cdot M$$

be the map induced by

$$1 \longrightarrow \tilde{e}_{i_0}.$$

This is well defined, since H acts on $\tilde{e}_{i_0}$ via $\chi_{i_0}^H$. By $\tilde{\pi}$ we denote the multiplication by π, and put $\overline{P} = P/(\ker(\tilde{\pi}))$, $\ker\tilde{\pi} = rad_R(P)$ and

$$\overline{M} = (M/\pi^t \cdot M)/\ker(\tilde{\pi}); \ker\tilde{\pi} = rad_R(M/\pi^t \cdot M).$$

Then we have the commutative diagram with exact rows and columns

$$
\begin{array}{ccccccccc}
0 & \longrightarrow & K_1 & \overset{\tilde{\pi}}{\longrightarrow} & K_2 & \longrightarrow & 0 \\
& & \downarrow & & \downarrow & & \downarrow \\
0 & \longrightarrow & ker\,\tilde{\pi} & \overset{\tilde{\pi}}{\longrightarrow} & P & \longrightarrow & \overline{P} & \longrightarrow & 0 \\
& & \downarrow \alpha' & & \downarrow \alpha & & \downarrow \overline{\alpha} \\
0 & \longrightarrow & ker\,\tilde{\pi} & \overset{\tilde{\pi}}{\longrightarrow} & M/\pi^t \cdot M & \longrightarrow & \overline{M} & \longrightarrow & 0
\end{array}
$$

Because of 2.8, $\overline{\alpha}$ is a split monomorphism. Moreover, P and $M/\pi^t \cdot M$ are $R/\pi^t \cdot R$–free on the same number of generators as $\overline{P}$ and $\overline{M}$ resp. Hence $Im(\alpha)$ is free on the number of generators of $\overline{P}$, which is the same as the number of free summands in P. Thus α and α' are monomorphisms. We then have the commutative diagram with exact rows

$$
\begin{array}{ccccccccc}
0 & \longrightarrow & P & \overset{\alpha}{\longrightarrow} & M/\pi^t \cdot M & \longrightarrow & C & \longrightarrow & 0 \\
& & \downarrow & & \downarrow & & \downarrow \\
0 & \longrightarrow & \overline{P} & \overset{\overline{\alpha}}{\longrightarrow} & \overline{M} & \longrightarrow & \overline{C} & \longrightarrow & 0
\end{array}
\;,
$$

where the vertical maps are reduction modulo π. The bottom row is split exact, and a splitting is given as follows (cf. 2.8): Recall that

$$
M/\pi^t \cdot M = \oplus_{i,U}((R/\pi^t \cdot R)_{\chi_i^U/\pi^t}) \uparrow_U^G \; .
$$

We now write $\tilde{e}_{i_0} = \Sigma\, v_{i,U}$ according to the above decomposition of $M/\pi^t \cdot M$. Then for $h \in H$ we have

$$
h \cdot \tilde{e}_{i_0} = (\chi_{i_0}^H/\pi^t)(h) \cdot \tilde{e}_{i_0}, \text{ and so}
$$

$$
\Sigma\, h \cdot v_{i,U} = \Sigma\, (\chi_{i_0}^H/\pi^t)(h) \cdot v_{i,U} \; .
$$

Since the $v_{i,U}$ lie in different RG–direct summands of $M/\pi^t \cdot M$, we conclude, that

$$
h \cdot v_{i,U} = (\chi_{i_0}^H/\pi^t)(h) \cdot v_{i,U};
$$

i.e. H acts on each $v_{i,U}$ via $\chi_{i_0}^H$.

Since by construction $Tr_{G/H}(\tilde{e}_{i_0}) \neq 0$, we conclude, that there exists a $v_{i_1,H}$, such that $Tr_{G/H}(v_{i_1,H}) \neq 0$. The map

$$
\overline{\beta} : \overline{M} = \oplus_{i,U}(\overline{R}_{\overline{\chi}_i^U}) \uparrow_U^G \longrightarrow \overline{P} \; .
$$

$(\overline{x}_{i,U}) \longrightarrow \overline{x}_{i_1,H}$ is then a splitting for $\overline{\alpha}$ modulo an automorphism of $\overline{P}$. By Nakayama's lemma it follows that the map

$$
\beta : M/\pi^t \cdot M = \oplus_{i,U}((R/\pi^t \cdot R)_{\chi_i^U/\pi^t}) \uparrow_U^G \longrightarrow P \; .
$$

$$
(x_{i,U}) \longrightarrow x_{i_1,H}
$$

is a splitting of α. This proves the claim. q.e.d.   □

We now define the RG–homomorphism

$$\varphi : R_{\chi_{i_0}^H} \uparrow_H^G \longrightarrow M \text{ by}$$

$$1 \longrightarrow e_{i_0}.$$

Applying the Ker-Coker-Lemma to the commutative diagram with exact rows

$$
\begin{array}{ccccccccc}
0 & \longrightarrow & R_{\chi_{i_0}^H} \uparrow_H^G & \xrightarrow{\pi} & R_{\chi_{i_0}^H} \uparrow_H^G & \longrightarrow & (R/\pi^t \cdot R)_{\chi_{i_0}^H} \uparrow_H^G & \longrightarrow & 0 \\
 & & \downarrow \varphi & & \downarrow \varphi & & \downarrow \tilde{\varphi} & & \\
0 & \longrightarrow & M & \xrightarrow{\pi} & M & \longrightarrow & M/\pi^t \cdot M & \longrightarrow & 0
\end{array},
$$

we conclude – using, that M and $R_{\chi_{i_0}^H} \uparrow_H^G$ are RG–lattices and applying Nakayama's lemma – that φ is a monomorphism and $\text{Coker}(\varphi)$ is an RG–lattice. In the commutative diagram

$$
\begin{array}{ccccccccc}
0 & \longrightarrow & R_{\chi_{i_0}^H} \uparrow_H^G & \xrightarrow{\varphi} & M & \longrightarrow & \text{Coker}(\varphi) & \longrightarrow & 0 \\
 & & \downarrow & & \downarrow & & \downarrow & & \\
0 & \longrightarrow & (R/\pi^t \cdot R)_{\chi_{i_0}^H} \uparrow_H^G & \xrightarrow{\tilde{\varphi}} & M/\pi^t \cdot M & \longrightarrow & \text{Coker}(\tilde{\varphi}) & \longrightarrow & 0
\end{array},
$$

the top sequence is an exact sequence of RG–lattices, the vertical map are reduction modulo π^t, and by the claim, the bottom row is split exact. Now an application of 2.1 shows, that the top row is split exact. By induction, $\text{Coker}(\varphi)$ is a generalized permutation lattice, and so M is a generalized permutation lattice.

This completes the proof of Theorem 1.3. q.e.d.   □

§3 The Proof of the Second Theorem

Recall the statement of Theorem 1.2:

Theorem. *Let N be normal in G , and let M be an RG–lattice, such that*

(a) *$M \downarrow_N$ is RN-free,*

(b) *$M/(I_R(N) \cdot M)$ is a G/N permutation module.*

Then M is a permutation module for G, as a matter of fact, $M \simeq \oplus RN \uparrow_{H_i}^G$, where $H_i \cap N = 1$.

PROOF: We shall do *several reductions*:

(34) *Reduction 1:* We may assume, that K is a splitting field for G and all of its subgroups.

PROOF: Let L with rings of integers S be a finite extension of K, such that L is a splitting field for G and all of its subgroups. Assume the theorem to be true for SG. Then

$$S \otimes_R M \simeq \oplus S\uparrow_{H_i}^{G} \simeq S \otimes_R (\oplus R\uparrow_{H_i}^{G}),$$

and the Noether-Deuring theorem [Ro; 72 2] implies

$$M \simeq \oplus R\uparrow_{H_i}^{G} .$$

q.e.d.   □

Hence we may assume that K is a splitting field for G and all of its subgroups.

(35) *Reduction 2:* We may assume that N is cyclic of order p.

PROOF: This reduction follows Weiss' argument [We; 87, Proof of Theorem 2]. Assume that the result is true for N cyclic of order p. We use induction on $|N|$. For $N = 1$ there is nothing to show. Choose C central in G of order p, generated by c and contained in N. Then $M/(c-1)\cdot M$ satisfies the hypotheses of the theorem for G/C and N/C, and hence is a permutation module; thus we are reduced to the case where N is cyclic of order p. q.e.d.   □

3.1. Pullbacks. We assume from now on, that $N = C = < c\,|\,c^p = 1 >$ is cyclic of order p, and that the quotient field of R, K is a splitting field for G and all of its subgroups, and that C is *central* in G. By ζ we denote a fixed primitive $p - th$ root of unity in R.

3.2. Notation. The central primitive idempotents in KC are given by

(36) $e_i = (1/p) \cdot \Sigma_{j=0}^{p-1}(\zeta^i)^j \cdot c^j,\ 0 \leq i \leq p - 1.$

(We keep this notation fixed for the remainder of this section; in particular, c and ζ are fixed.)

(37) $\eta_i = \Sigma_{j=0}^{i} e_j\ ;\ 0 \leq i \leq p - 1.$

Since C is central in G, the idempotents $\{e_i\}_{0 \leq i \leq p-1}$ are central in KG.

We shall next describe the pullback structure of $RC\eta_i$:

3.3. Lemma. *We have a pullback*

$$
\begin{array}{ccc}
RC \cdot \eta_{i+1} & \xrightarrow{\lambda_{i+1}} & RC \cdot \eta_i \\
\downarrow \kappa_{i+1} & & \downarrow \mu_i \\
RC \cdot e_{i+1} & \xrightarrow{\nu_{i+1}} & \hat{\Lambda}_i
\end{array},
$$

where we have the following description of the kernels:

$$Ker(\lambda_{i+1}) = \cap_{j=0}^{i}(c - \zeta^j) \cdot RC \cdot \eta_{i+1},$$

$$Ker(\nu_{i+1}) = \cap_{j=0}^{i}(\zeta^{i+1} - \zeta^j)RC \cdot e_{i+1} = (1 - \zeta)^{i+1} \cdot R,$$

$$Ker(\kappa_{i+1}) = (c - \zeta^{i+1}) \cdot RC \cdot \eta_{i+1}, Ker(\mu_i) = (c - \zeta^{i+1}) \cdot RC \cdot \eta_i.$$

In addition,

$$\hat{\Lambda}_i \simeq R/(1 - \zeta)^{i+1} \cdot R,$$

on which c acts as ζ^{i+1}, where the action comes from $RC \cdot e_{i+1}$, on the other hand it acts in the same way, when the action is induced from $RC \cdot \eta_i$.

PROOF: This will be done in *several steps*. We first note:

(38) Let Λ be an $R-$order in a semi simple $K-$algebra A, and let $1 = \varepsilon_1 + \varepsilon_2$ be a decomposition of the identity into orthogonal central idempotents. Then Λ is a pullback

$$
\begin{array}{ccc}
\Lambda & \xrightarrow{pr_1} & \Lambda \cdot \epsilon_1 \\
\downarrow pr_2 & & \downarrow \\
\Lambda \cdot \epsilon_2 & \longrightarrow & \hat{\Lambda}
\end{array}
,
$$

where $\hat{\Lambda} = \Lambda/((\Lambda \cdot \varepsilon_1 \cap \Lambda) \oplus (\Lambda \cdot \varepsilon_2 \cap \Lambda))$ is isomorphic to $\Lambda \cdot \varepsilon_i/(\Lambda \cap \Lambda \cdot \varepsilon_i)$, $i = 1, 2$.

PROOF: We have a commutative diagram with exact rows and columns from which the statement follows:

$$
\begin{array}{ccccccccc}
 & & & & 0 & & 0 & & \\
 & & & & \downarrow & & \downarrow & & \\
 & & & & \Lambda \cap \Lambda e_1 & = & \Lambda \cap \Lambda e_1 & & \\
 & & & & \downarrow & & \downarrow & & \\
0 & \longrightarrow & \Lambda \cap \Lambda e_2 & \longrightarrow & \Lambda & \longrightarrow & \Lambda e_1 & \longrightarrow & 0 \\
 & & \| & & \downarrow & & \downarrow & & \\
0 & \longrightarrow & \Lambda \cap \Lambda e_2 & \longrightarrow & \Lambda e_2 & \longrightarrow & \hat{\Lambda} & \longrightarrow & 0 \\
 & & & & \downarrow & & \downarrow & & \\
 & & & & 0 & & 0 & &
\end{array}
$$

$\square$

(39) The kernel of the map

$$\kappa_{i+1} : RC \cdot \eta_{i+1} \longrightarrow RC \cdot e_{i+1}$$

is $Ker(\kappa_{i+1}) = (c - \zeta^{i+1}) \cdot RC \cdot \eta_{i+1}$. It should be noted, that $(c - \zeta^{i+1})$ is a central element in RG; however, it is a zero divisor, even in $RC \cdot \eta_{i+1}$.

PROOF: We have the commutative diagram with exact rows

$$
\begin{array}{ccccccccc}
0 & \longrightarrow & (c - \zeta^{i+1}) \cdot RC & \longrightarrow & RC & \longrightarrow & RC \cdot e_{i+1} & \longrightarrow & 0 \\
 & & \downarrow \varphi' & & \downarrow \varphi & & \| & & \\
0 & \longrightarrow & Ker(\kappa_{i+1}) & \longrightarrow & RC \cdot \eta_{i+1} & \longrightarrow & RC \cdot e_{i+1} & \longrightarrow & 0
\end{array}
$$

Since φ' and φ are surjective, we conclude, that $Ker(\kappa_{i+1})$ is the image of the restriction of φ – *i.e.* multiplication with η_{i+1} to $(c - \zeta^{i+1}) \cdot RC$; but this is the statement of the claim. □

(40) We have
$$Ker(\lambda_{i+1}) = \cap_{j=0}^{i}(c - \zeta^j) \cdot RC \cdot \eta_{i+1}.$$

PROOF: The kernel of λ_{i+1} is equal to the kernel of the projections:
$$RC \cdot \eta_{i+1} \longrightarrow \Pi_{j=0}^{i} RC \cdot e_j.$$

Because of equation (39) this is the intersection of the kernels of the various projections, as claimed. □

We have

(41)
$$\begin{aligned} Ker(\nu_{i+1}) &= \Pi_{j=0}^{i}(\zeta^{i+1} - \zeta^i) \cdot RC \cdot e_{i+1}. \\ Ker(\mu_i) &= (c - \zeta^{i+1}) \cdot RC \cdot \eta_i; \end{aligned}$$

Note that on $\Lambda \cdot \eta_i$, the element $(c - \zeta^{i+1})$ is not a zero divisor.

PROOF: Because of the pullback diagram 3.3, we know that
$$Ker(\nu_{i+1}) \simeq Ker(\lambda_{i+1})$$

via κ_{i+1} and
$$Ker(\mu_i) \simeq Ker(\kappa_{i+1})$$

via λ_{i+1}.

As for the kernel of ν_{i+1}, we note that $RC \cdot e_{i+1} \simeq R$ is a principal ideal domain, on which c acts as ζ^{i+1}. This proves (41). □

We then conclude:
$$\hat{\Lambda}_i \simeq R/(\prod_{j=1}^{i}(\zeta^{i+1} - \zeta^j)R),$$

since
$$\hat{\Lambda}_i \simeq \Lambda \cdot e_{i+1}/Ker(\nu_{i+1}).$$

This proves 3.3. q.e.d. □

3.4. Lemma. *We have a commutative diagram*

$$\begin{array}{ccc} RC \cdot \eta_{i+1} & \xrightarrow{\lambda_{i+1}} & RC \cdot \eta_i \\ \downarrow \mu_{i+1} & & \downarrow \mu_1 \\ \hat{\Lambda}_{i+1} & \xrightarrow{\sigma_{i+1}} & \hat{\Lambda}_i \end{array} \quad .$$

In order to define $\sigma_{i+1} : \hat{\Lambda}_{i+1} \longrightarrow \hat{\Lambda}_i$, *we identify – according to 3.3 –*

$$\hat{\Lambda}_{i+1} \cong R/(\prod_{j=0}^{i+1}(\zeta^{i+2} - \zeta^j) \cdot R =: \hat{R}_{i+1},$$

on which c *acts as* ζ^{i+2} *and*

$$\hat{\Lambda}_i \cong R/(\prod_{j=0}^{i}(\zeta^{i+1} - \zeta^j) \cdot R =: \hat{R}_i,$$

on which c *acts as* ζ^{i+1}. *Then* σ_{i+1} *is reduction modulo*

$$(\prod_{j=0}^{i}(\zeta^{i+1} - \zeta^j))/(\prod_{j=0}^{i+1}(\zeta^{i+2} - \zeta^j)) =: \hat{J}.$$

PROOF: According to 3.3 we have the commutative diagram with exact rows and columns: (For abbreviation we denote for the next diagram $X_i := RC \cdot \eta_i$ and $Y_i := RC \cdot e_{i+1}$.)

$$
\begin{array}{ccccccccc}
0 & \longrightarrow & \hat{J}_i & \xrightarrow{\alpha} & \hat{R}_{i+1} & \xrightarrow{\sigma_{i+1}} & \hat{R}_i & \longrightarrow & 0 \\
 & & \uparrow & & \uparrow \mu_{i+1} & & \uparrow \mu_i & & \\
0 & \longrightarrow & \prod_{j=0}^{i}(\zeta^{j+1} - \zeta^j)Y_{i+1} & \longrightarrow & X_{i+1} & \xrightarrow{\lambda_{i+1}} & X_i & \longrightarrow & 0 \\
 & & \uparrow \beta & & \uparrow \gamma & & \uparrow \delta & & \\
0 & \longrightarrow & \prod_{j=0}^{i+1}(\zeta^{j+2} - \zeta^j)Y_{i+1} & \longrightarrow & (c - \zeta^{i+2})X_{i+1} & \longrightarrow & (c - \zeta^{i+1})X_i & \longrightarrow & 0
\end{array}
$$

In fact, we note that $RC \cdot e_{i+1} = R$, where c acts via ζ^{i+1}. Since γ and δ are injective, the kernel of λ_{i+1} restricted to the kernel of μ_{i+1} is $(c - \zeta^{i+2}) \cdot Ker(\lambda_{i+1})$. However, c acts as ζ^{i+1} on $Ker(\lambda_{i+1})$, whence this kernel is as we have described it. Moreover, the cokernel of β is $\hat{J}$. Thus $\hat{J}$ is the kernel of σ_{i+1}. Moreover, since under σ_{i+1} the element ζ^{i+2} is identified with ζ^{i+1}, the map σ_{i+1} is a G–homomorphism.

This proves 3.4. q.e.d.   □

Permutation modules

We shall use the following

3.5. Definition. We shall call a permutation module

$$M = \oplus_{i=1}^{t} R \uparrow_{H_i}^{G}$$

a C–*free* RG-*permutation lattice, provided* $H_i \cap C = 1$.

3.6. Claim. *Let* $M = R\uparrow_H^G$ *be a transitive* C–*free permutation lattice. Then*

$$R\uparrow_H^G \cong RC \uparrow_{H \times C}^{G},$$

where H acts trivially on RC. Moreover,

$$End_{RG}(R\uparrow_H^G) \simeq Hom_{RC}(RC, RC)^{(t)}$$

for some $t \in I\!\!N$.

PROOF: We can choose coset representatives through C. Now C lies in the centre of G, and so

$$R\uparrow_H^{H \times C} \simeq RC|_{H \times C},$$

Here C acts regularly on RC and H acts trivially on RC. Thus we obtain

$$R\uparrow_H^G \simeq RC|_{H \times C} \otimes_{R(H \times C)} RG.$$

In order to compute the RG−endomorphisms of $R\uparrow_H^G$, we let more generally X be an RC−lattice, on which the subgroup H acts trivially. Then by Mackey's theorem [C-R1; 82, (10.13)]

$$X\uparrow_{H \times C}^G \downarrow_{H \times C} \simeq \oplus X \downarrow_{({}^g H \cap H) \times C}\uparrow^{H \times C},$$

where the sum is taken over the double cosets of $H \times C$ in G.

Using Frobenius reciprocity, we get the following chain of natural isomorphisms:

$$\begin{aligned} End_{RG}(X\uparrow_{H \times C}^G) &\cong Hom_{R(H \times C)}(X, \oplus X\uparrow_{({}^g H \cap H) \times C}^{H \times C}) \\ &\cong Hom_{R(({}^g H \cap H) \times C)}(X, \oplus X) \cong End_{RC}(X, X)^{(t)}, \end{aligned}$$

since $({}^g H \cap H)$ acts trivially on X. This completes the proof of 3.6.   □

The above arguments also prove:

3.7.

Let $M = R\uparrow_H^G$ be a transitive C−free permutation lattice, and let e be a central idempotent in KC, then

$$M \cdot e \simeq RC \cdot e|_{H \times C} \otimes_{R(H \times C)} RG,$$

and

$$End_{RG}(M \cdot e) \simeq End_{RC \cdot e}(RC \cdot e)^{(t)}.$$

3.8. Lemma. *Let $M = \oplus R\uparrow_{H_i}^G$ be a C−free permutation module. Then with the above notation, for $0 \le i \le p - 2$, the induced map*

$$End_{RG}(M \cdot \eta_i) \longrightarrow End_{RG}(M \cdot \eta_i/(c - \zeta^{i+1}) \cdot M \cdot \eta_i)$$

is surjective. (Note that η_i is central, and so $M \cdot \eta_i$ is to be understood as $\eta_i \cdot M$.)

PROOF: (1) In case M is a transitive C-free permutation module, the statement follows from the description of the endomorphisms in (3.7).

(2) The lemma will follow, if we can show:

Let $M_i = R \uparrow_{H_i}^G$, $i = 1, 2$, be transitive C-free permutation modules, such that

$$M_1 \cdot \eta_i / ((c - \zeta^{i+1}) \cdot M_1 \cdot \eta_i) \simeq M_2 \cdot \eta_i / ((c - \zeta^{i+1}) \cdot M_2 \cdot \eta_i),$$

then $M_1 \cdot \eta_i \simeq M_2 \cdot \eta_i$, $0 \le i \le p - 2$. To do so, we shall be using induction on i :

$i = 0$: We recall, that $\eta_0 = e_0$ is the trivial idempotent (equations (36) and (37)), and so $M_1 \cdot \eta_0 = R \uparrow_{H_1}^{G/C}$ and $M_2 \cdot \eta_0 = R \uparrow_{H_2}^{G/C}$ are genuine permutation modules for G/C ; moreover, since c acts trivially on η_0, the reduction modulo $(c - \zeta^{0+1})$ is just reduction modulo $(1 - \zeta)$, and so

$$M_1 \cdot \eta_0 \simeq M_2 \cdot \eta_0 \text{ iff } M_1 \cdot \eta_0 / (1 - \zeta) \cdot M_1 \cdot \eta_0 \simeq M_2 \cdot \eta_0 / (1 - \zeta) \cdot M_2 \cdot \eta_0.$$

This proves the case $i = 0$.

$i > 0$: Assume that the modules $M_j \cdot \eta_i / ((c - \zeta^{i+1}) \cdot M_j \cdot \eta_i)$ are isomorphic for $j = 1, 2$. We recall from 3.3, that

$$RC \cdot \eta_i / (c - \zeta^{i+1}) \cdot RC \cdot \eta_i \simeq R / (1 - \zeta)^{i+1} R \simeq RC \cdot e_{i+1} / (1 - \zeta)^{i+1} RC \cdot e_{i+1}$$

(note $i < p - 1$), and c acts as ζ^{i+1} on these modules. In particular, we have an isomorphism of RG-modules

$$M_j \cdot \eta_i / ((c - \zeta^{i+1}) \cdot M_j \cdot \eta_i) \simeq M_j \cdot e_{i+1} / (1 - \zeta)^{i+1} M \cdot e_{i+1} =: \hat{M}_{j,i}, \ j = 1, 2.$$

We now invoke lemma 3.4, which implies that we have a commutative diagram of RG-modules

$$
\begin{array}{ccc}
M_j \cdot \eta_i & \xrightarrow{\lambda_i} & M_j \cdot \eta_{i-1} \\
\downarrow \mu_i & & \downarrow \mu_{i-1} \\
\hat{M}_{j,i} & \xrightarrow{\sigma_{ji}} & \hat{M}_{j,i-1}
\end{array} \quad , \ j = 1, 2.
$$

The RG-isomorphism $\hat{M}_{1,i} \simeq \hat{M}_{2,i}$, which we are given by assumption, now induces an RG-isomorphism $\hat{M}_{1,i-1} \simeq \hat{M}_{2,i-1}$, and by induction hypothesis, the modules $M_1 \cdot \eta_{i-1}$ and $M_2 \cdot \eta_{i-1}$ are isomorphic.

We shall show, that this together with the assumption implies that $M_1 \cdot \eta_i$ and $M_2 \cdot \eta_i$ are isomorphic. In fact we shall *show even more*:

3.9. Claim. *Assume that $\hat{M}_{1,1}$ and $\hat{M}_{2,1}$ are isomorphic as RG-modules, then M_1 and M_2 are already isomorphic.*

PROOF: Recall, that $M_j = R \uparrow_{H_j}^G$, $j = 1, 2$ are C-free permutation modules. The arguments above show, that $\hat{M}_{1,0}$ and $\hat{M}_{2,0}$ are isomorphic. The case $i = 0$

then shows, that $M_1 \cdot e_0$ and $M_2 \cdot e_0$ are isomorphic; i.e. $R \uparrow_{H_1}^{G/C}$ and $R \uparrow_{H_2}^{G/C}$ are isomorphic. Since the groups involved are p−groups, we conclude that H_1 and H_2 are conjugate in G/C [Con; 72]. Replacing H_1 by a G−conjugate, we may assume that $H_1 = H_2$ in G/C; i.e. $C \times H_1 = C \times H_2$. Recall from 3.3

$$\hat{M}_{j,1} \simeq R/(1-\zeta)^2 \cdot R \otimes_{R(C \times H_j)} RG \ , j = 1, 2,$$

and the vertices of $\hat{M}_{1,1}$ and $\hat{M}_{2,1}$ coincide and are equal to $C \times H_1 = C \times H_2$; Moreover, the sources [23] of both modules, $R/(1-\zeta)^2 \cdot R$ as $C \times H_1$ and $C \times H_2$ modules – recall that c acts as ζ^2 on $R/(1-\zeta)^2 \cdot R$, and H_1 and H_2 act trivially – must be isomorphic. However, every $h_1 \in H_1$ can be written as $h_2 \cdot c^{\varrho(h_1)} = h_1$ for some $h_2 \in H_2$. But since the action of h_2 on $R/(1-\zeta)^2 \cdot R$ is trivial, h_1 can only act trivially if $\varrho(h_1) \equiv 0 \ mod(p)$, since this is the only case, where $\zeta^{\varrho(h_1)}$ acts trivially on $R/(1 - \zeta)^2 \cdot R$. But then $H_1 = H_2$ and even M_1 and M_2 are isomorphic.

This proves Claim 3.9   □

and also lemma 3.8   □

   We are now in the position to prove the Theorem 1.2
in case $N = C$ is cyclic of order p and R a splitting ring for G :

   Let M be an RG-lattice, such that

(a) M as RC-module is RC-free,

(b) $M \cdot e_0$ is an $R(G/C)$−permutation module (e_0 the trivial idempotent of RC, see (36)).

The *aim* is to show that M is an RG−permutation module. We observe, that the question of whether a module is a pullback, is a purely set theoretical one, and since M as RC−module is RC−free, we get similar pullbacks for $M \cdot \eta_i$− recall that $M \cdot \eta_i = \eta_i \cdot M$, since η_i is central in RG – as we have derived for RC (cf. 3.3, 3.4). The *strategy* is as follows:

   We show inductively, that

$$M \cdot \eta_i \simeq \oplus_{j=1}^{t} R \uparrow_{H_j}^{G} \ \text{ with } H_j \cap C = 1.$$

$i = 0$: The condition (b) implies that

$$M \cdot \eta_0 \simeq \oplus_{j=1}^{t} R \uparrow_{\hat{H}_j}^{G/C} \ , \text{ where } \hat{H}_j \text{ is a subgroup of } G/C, 1 \le j \le t.$$

Let $\tilde{H}_j$ be preimages in G, then

$$M \cdot \eta_0 \simeq \oplus_{j=1}^{t} R \uparrow_{\tilde{H}_j}^{G} \ .$$

---

[23]   If V is a vertex of the indecomposable RG-lattice M, say $M|N \uparrow_{V}^{G}$, for some indecomposable RV−lattice N, then N is called a source.

The case $i = 1$: Using (a) and the remark above, we have a pullback diagram (cf. 3.4)

$$
\begin{array}{ccc}
& & (1-\zeta)Me_0 \\
& & \downarrow \\
M \cdot \eta_1 & \longrightarrow & M \cdot \eta_0 \\
\downarrow & & \downarrow \\
0 \longrightarrow (\zeta-1)Me_1 \longrightarrow Me_1 & \longrightarrow & Me_0/(1-\zeta)Me_0 \longrightarrow 0
\end{array}
$$

Now $Me_0/(1-\zeta)M_0$ is a permutation module, and so we can invoke Theorem 1.3 to conclude that $M \cdot e_1$ is a generalized permutation module, say

$$
Me_1 \simeq \oplus_{j=1}^{t} R_j \uparrow_{\tilde{H}_j}^{G},
$$

where $\tilde{H}_j \longrightarrow R_j$ is an abelian character of $\tilde{H}_j$. However, c acts as ζ on $M \cdot e_1$ and $\tilde{H}_j/C$ acts trivially – the case $i = 0$– and so $\tilde{H}_j = H_j \times C$ for a subgroup H_j, which acts trivially on R_j.

We now consider the permutation module

$$
M_0 := \oplus_{j=1}^{t} RC \downarrow_{H_j \times C} \uparrow^{G},
$$

and we claim, that

$$
M \cdot \eta_1 \simeq M_0 \cdot \eta_1.
$$

But this is true, since both modules have isomorphic projections with respect to η_0 and e_1, and since endomorphisms of

$$
M_0 \cdot e_0/(1-\zeta)M_0 \cdot e_0
$$

can be lifted to endomorphisms of $M \cdot e_0$, thanks to 3.8. Thus both pullbacks, the one above and the analogous one for $M_0\eta_1$, coincide and so the claim follows.

The general case $i > 0$: By induction,

$$
M \cdot \eta_{i-1} \simeq M_0 \cdot \eta_{i-1}.
$$

Using the pullback diagram in Lemma 3.3 for M and Theorem 1.3, we conclude that $M \cdot e_i$ is a generalized permutation module, which must be isomorphic to $M_0 \cdot e_i$ (cf. the case $i = 1$). But then $M_0 \cdot \eta_i$ and $M \cdot \eta_i$ have isomorphic projections with respect to η_{i-1} and e_i and so we can apply Lemma 3.8 to conclude that $M_0 \cdot \eta_i \simeq M \cdot \eta_i$.

Eventually we reach $i = p-1$, and the statement follows. This concludes the proof of Theorem 1.2. q.e.d.   □

VII Global units

§1 Picard–groups and automorphisms

In this chapter we will introduce and extend concepts due to Fröhlich, the Picard group [Fr; 73], which shall help us to study automorphism groups of orders. These Picard–groups are invariant under Morita equivalence. There is a map from the automorphism group of orders to their Picard group. The kernel of this map is the group of inner automorphisms.

We follow the work of Fröhlich and refer also to [Re; 75] and [C-R2; 87].

Let us go into more detail: Throughout this paragraph Λ denotes an R – order in a finite dimensional separable $K = frac(R)$–algebra A, where K is assumed to be an algebraic number field and R a Dedekind domain.

1.1. Definition. *An invertible bimodule for an order Λ is a (Λ, Λ) – bimodule $_\Lambda M_\Lambda$ with the following additional properties:*

(a) *There exists a (Λ, Λ) – bimodule $_\Lambda N_\Lambda$ with isomorphisms*

(b) $N \otimes M \xrightarrow{\alpha} \Lambda \xleftarrow{\beta} M \otimes N$ *as bimodules*

(c) *such that the following diagrams commute:*

$$
\begin{array}{ccc}
M \otimes N \otimes M & \xrightarrow{\beta \otimes 1} & \Lambda \otimes M \\
\downarrow 1 \otimes \alpha & & \downarrow \\
M \otimes \Lambda & \longrightarrow & M
\end{array}
\quad , \quad
\begin{array}{ccc}
N \otimes M \otimes N & \xrightarrow{\alpha \otimes 1} & \Lambda \otimes N \\
\downarrow 1 \otimes \beta & & \downarrow \\
N \otimes \Lambda & \longrightarrow & N
\end{array} \quad .
$$

We see that the definition above ensures that the set of isomorphism classes of invertible bimodules forms a group under the formation of tensor products. The inverse of a bimodule class (M) is the bimodule class (N) as in the definition, the unit element is $(_\Lambda \Lambda_\Lambda)$.

Moreover, Morita theory tells us that the set of isomorphism classes of invertible bimodules is the group of auto equivalences of the module category Mod_Λ of Λ–modules.

1.2. Definition. *The S-Picard-group $Pic_S\Lambda$ of an R - order Λ over a subring $S \supseteq R$ of the center of Λ is the set of isomorphism classes (M) of invertible (Λ, Λ)-bimodules M such that S acts the same on both sides (i.e. for all $m \in M, s \in S$ is $s \cdot m = m \cdot s$). We also call those bimodules M 'bimodules over S '.*

We give some examples.

1.) Let R be the ring of algebraic integers in an algebraic number field K. If $\mathcal{A}$ is an ideal in R, then it is an R–bimodule over $\mathbb{Z}$, but also an R–bimodule over R. Moreover, $\mathcal{A}$ is invertible, since R is a Dedekind domain.

2.) Let R be as in 1.) and assume in addition, that R is Galois over $\mathbb{Z}$. For $\alpha \in Gal(R/\mathbb{Z})$ the twisted module (cf. 3.) $_\alpha R_1$ is an R–bimodule over

R_0, if R_0 is the fixed ring of α. In particular $_\alpha R_1$ is always an R-bimodule over $\mathbb{Z}$ (cf. below).

3.) Let α and β be automorphisms of the order Λ. Then we form the twisted bimodule $_\alpha\Lambda_\beta$, which is Λ as abelian group, but the action of Λ is twisted by α and β: For $r, s \in \Lambda$ and $m \in {_\alpha\Lambda_\beta}$ we define

$$r \bullet m \bullet s := \alpha(r) \cdot m \cdot \beta(s).$$

This module is invertible. This follows from the formulae in 1.3 below.

These bimodules are of particular interest in connexion with the isomorphism problem, since automorphisms are interpreted as modules, for which a variety of constructions are available.

So the study of Picard groups includes the study of automorphisms and the class group.

1.3. Lemma. *For automorphisms α and β of the R-order Λ we have*

(a) $_\alpha\Lambda_\beta \simeq {_{\beta^{-1}\alpha}\Lambda_1}$,

(b) $_\alpha\Lambda_1 \otimes_\Lambda {_1\Lambda_\beta} \simeq {_\alpha\Lambda_\beta}$,

(c) $_\gamma\Lambda_1 \simeq {_1\Lambda_1}$ *for every inner automorphism γ.*

PROOF : We note that

$$\beta^{-1} : {_\alpha\Lambda_\beta} \longrightarrow {_{\beta^{-1}\alpha}\Lambda_1}$$

is Λ – linear on both sides:

$$\beta^{-1}(r \bullet m \bullet s) = \beta^{-1}(\alpha(r)m\beta(s)) = \beta^{-1}\alpha(r)\beta^{-1}(m)s = r \bullet \beta^{-1}(m)s$$

for all $r, s, m \in \Lambda$. For the second formula, the map induced from $r \otimes s \longrightarrow rs$ gives the isomorphism.

The third equation comes from multiplication with the conjugating element:

$$c^{-1}(\gamma(r)m) = c^{-1}(crc^{-1}m) = r(c^{-1}m)$$

and we get the desired isomorphism. q.e.d.   □

Using the lemma from above we can define a homomorphism

$$\Omega : Aut_S(\Lambda) \longrightarrow Pic_S(\Lambda)$$

from the automorphism group of Λ fixing the subring S of the center of Λ elementwise to the Picard-group of Λ over S if we define

$$\Omega(\alpha) := ({_\alpha\Lambda_1})$$

Let us discuss the kernel:

If $\Omega(\alpha) = ({}_1\Lambda_1)$ then there is an isomorphism $\phi : {}_\alpha\Lambda_1 \longrightarrow \Lambda$ of bimodules. For all $r \in \Lambda$ we have

$$\phi(r) = \phi(r \bullet 1) = r \bullet \phi(1) = \alpha(r)\phi(1) = \phi(1 \bullet r) = \phi(1) \bullet r = \phi(1)r.$$

Since ϕ is bijective, $\phi(1)$ is invertible. So the calculation above yields

$$\alpha(r) = \phi(1)r\phi(1)^{-1} \text{ for all } r \in \Lambda$$

and our automorphism α is inner. We thus have an exact sequence of groups:

$$1 \longrightarrow Inn(\Lambda) \longrightarrow Aut_S(\Lambda) \longrightarrow Pic_S(\Lambda)$$

The following observation is now immediate: In general the Picard–group is not abelian, since the outer automorphism group $Out_S(\Lambda)$ is in general not abelian.

An obvious but important property of Picard groups is that they do only depend on the Morita equivalence class of an R–order Λ.

1.4. Proposition. *If the R–algebras Λ and Δ are Morita–equivalent and $R \subseteq S \subseteq cent(\Lambda)$, then*

$$Pic_S(\Lambda) \simeq Pic_S(\Delta)$$

PROOF: Let ${}_\Lambda M_\Delta$ be an invertible (Λ, Δ)– bimodule with inverse ${}_\Delta N_\Lambda$ inducing the Morita correspondence. Then the map $(X) \longrightarrow (N \otimes_\Lambda X \otimes_\Lambda M)$ gives an isomorphism $Pic_R(\Lambda) \longrightarrow Pic_R(\Delta)$. Note that this is just a conjugation. $\square$

We state without proof, since a generalization will be proved later on, one of the most important properties of Picard groups.

Let $Picent(\Lambda) := Pic_{cent(\Lambda)}(\Lambda)$.

1.5. Theorem. *[Fr; 73][Re; 75, 37.28] Let Λ be an R–order in a separable K–algebra A, where Λ is a Dedekind domain with quotient field K and let $C = cent(\Lambda)$ be the center of Λ. For each maximal ideal $\wp$ of R, let $\Lambda_\wp$ be the $\wp$–adic completion of Λ. Then $Picent(\Lambda_\wp) = 1$ almost everywhere [24], and there is an exact sequence*

$$1 \longrightarrow Picent(C) \longrightarrow Picent(\Lambda) \longrightarrow \prod_\wp Picent(\Lambda_\wp) \longrightarrow 1.$$

§2   Conjugacy Classes of Finite Subgroups in $V(\mathbb{Z}G)$

In this section a theory closely related to the theory of invertible bimodules by Fröhlich (paragraph 1) is developed in order to describe conjugacy classes of

---

[24]   In fact, $\Lambda_\wp$ is separable for almost all primes $\wp$, and so it is a product $\Lambda_\wp = \prod_i (R_{\wp_i})_{n_i}$, where $R_{\wp_i}$ are unramified extensions of R.

finite subgroups in $V(\mathbb{Z}G)$. We shall apply these results in paragraph 3 to the symmetric group on three letters to obtain the results of Hughes and Pearson [H-P; 72] and to the dihedral groups of order 8 and 16.

Let G be a finite group, and let H_0 be a fixed subgroup of G. If

$$\alpha : H_0 \longrightarrow U(\mathbb{Z}G)$$

is a homomorphism, then we may view $\mathbb{Z}G$ as a $(H_0 \times G^{op})$–bimodule via α — denoted by ${}_\alpha \mathbb{Z}G_G$:

$$(u \times g) \cdot x := \alpha(u) \cdot x \cdot g, \, u \in H_0, \, g \in G, \, x \in \mathbb{Z}G.$$

If $H_0 = G$, then ${}_\alpha\mathbb{Z}G_G$ is an invertible bimodule (cf. the map Ω in paragraph 1), and we shall next translate some of the results on invertible bimodules to the present situation:

2.1. Lemma. *Let M be an $(H_0 \times G^{op})$-bimodule. M is of the form ${}_\alpha\mathbb{Z}G_G$ for some α as above iff $M|_{\mathbb{Z}G}$ is isomorphic to $\mathbb{Z}G$ as right $\mathbb{Z}G$-module.*

PROOF: Let M be right free of rank one; i.e. $M = x \cdot \mathbb{Z}G$ and so

$$h \cdot x = x \cdot \alpha(h), \, h \in H_0 \text{ for some } \alpha(h) \in U(\mathbb{Z}G).$$

Then the map $\alpha : H_0 \longrightarrow U(\mathbb{Z}G)$ is the desired homomorphism as is easily seen. The converse is obvious. q.e.d. □

Among the augmented automorphisms of $\mathbb{Z}G$, the central automorphisms $Autcent(\mathbb{Z}G)$ play an important role:

For such an automorphism α, the image of G is of the form $a \cdot G \cdot a^{-1}$ for some unit a in $\mathbb{Q}G$, which normalizes $\mathbb{Z}G$. In fact, α induces an automorphism – also denoted by α– on $\mathbb{Q}G$, which leaves the center of the semi simple $\mathbb{Q}$–algebra $\mathbb{Q}G$ elementwise fixed, and hence by the theorem of Skolem Noether [C-R1; 82] α as homomorphism of $\mathbb{Q}G$ is conjugation with a unit a in $\mathbb{Q}G$.

In view of the Zassenhaus conjecture for subgroups (section VI), those $(H_0 \times G^{op})$–bimodules M are of particular importance, where the action of H_0 on M is given by left multiplication with $a \cdot H_0 \cdot a^{-1} \leq U(\mathbb{Z}G)$ for some $a \in U(\mathbb{Q}G)$; with other words the subgroups H of $U(\mathbb{Z}G)$, which are of the form $a \cdot H_0 \cdot a^{-1}$ for some unit a in $\mathbb{Q}G$.

2.2. Remark. *The element a from above need not normalize $\mathbb{Z}G$; if it does normalize $\mathbb{Z}G$, then $a \cdot H_0 \cdot a^{-1}$ is part of the group basis $a \cdot G \cdot a^{-1}$ and conversely.*

Note that in this case automatically $U \leq V(\mathbb{Z}G)$. We characterize these bimodules in the following way:

2.3. Lemma. *Let M be an $(H_0 \times G^{op})$-bimodule. M is isomorphic to a bimodule of the form ${}_{a \cdot H_0 \cdot a^{-1}}\mathbb{Z}G_G$, for a unit $a \in \mathbb{Q}G$ and $a \cdot H_0 \cdot a^{-1} = U \leq V(\mathbb{Z}G)$ iff*

(a) $M|_{\mathbb{Z}G}$ is right $\mathbb{Z}G$-free of rank one,

(b) $\mathbb{Q}M$ is isomorphic as $(H_0 \times G^{op})$-bimodule to $_{H_0}\mathbb{Q}G_G$.

PROOF: Because of (a), $M = x \cdot \mathbb{Z}G$ as bimodule. (b) implies that there is a bimodule isomorphism $x \cdot \mathbb{Q}G \longrightarrow \mathbb{Q}G$, sending x to a unit a in $\mathbb{Q}G$, which maps $x \cdot \mathbb{Z}G$ to $a \cdot \mathbb{Z}G$. Now the statement follows directly; the converse statement is obvious. q.e.d.   □

2.4. Remark. The unit a in $\mathbb{Q}G/\mathcal{Z}(\mathbb{Q}G)$ ($\mathcal{Z}(\mathbb{Q}G)$ is the centre of $\mathbb{Q}G$) is by (a) and (b) only uniquely determined up to multiplication with a unit in $V(\mathbb{Z}G)$. More precisely, the bimodule class of $_{a \cdot H_0 \cdot a^{-1}}\mathbb{Z}G_G$ determines $a \cdot H_0 \cdot a^{-1}$ only up to conjugation with a unit in $V(\mathbb{Z}G)$. In fact assume that

$$\sigma : \ _{a \cdot H_0 \cdot a^{-1}}\mathbb{Z}G_G \longrightarrow \ _{H_0}\mathbb{Z}G_G$$

is a bimodule isomorphism, then we put $u = \sigma(1)$ and conclude, that u must be a unit in $\mathbb{Z}G$ with

$$h \cdot u = \sigma(a \cdot h \cdot a^{-1} \cdot 1) = \sigma(1 \cdot a \cdot h \cdot a^{-1}) = u \cdot a \cdot h \cdot a^{-1} \text{ for all } h \in H.$$

Thus $u^{-1} \cdot h \cdot u = a \cdot h \cdot a^{-1}$. Hence the isomorphism classes of $(H_0 \times G^{op})$-bimodules satisfying (a) and (b) classify the $V(\mathbb{Z}G)$-conjugacy classes of the subgroups $H \leq V(\mathbb{Z}G)$, such that there exists a unit $a \in \mathbb{Q}G$ with $a \cdot H \cdot a^{-1} = H_0$.

In order to develop a theory similar to that one of central invertible bimodules for the above $H_0 \times G^{op}$-bimodules, we make the following definitions (cf. [Fr]).

2.5. Definition. Let H_0 be a fixed subgroup of G. Then

- $Pic(H_0, \mathbb{Z}G)$ is the set of isomorphism classes of $(H_0 \times G^{op})$-bimodules M with $M|_{\mathbb{Z}G}$ projective of rank one.

- $Picent(H_0, \mathbb{Z}G)$ is the set of isomorphism classes of $(H_0 \times G^{op})$-bimodules M with $M|_{\mathbb{Z}G}$ projective of rank one and with $\mathbb{Q}M \simeq \mathbb{Q}G$ as $(H_0 \times G^{op})$-bimodule.

- $Out(H_0, \mathbb{Z}G)$ is the set of isomorphism classes of $(H_0 \times G^{op})$-bimodules M with $M|_{\mathbb{Z}G}$ free of rank one.

- $Outcent(H_0, \mathbb{Z}G)$ is the set of isomorphism classes of $(H_0 \times G^{op})$-bimodules M with $M|_{\mathbb{Z}G}$ free of rank one and with $\mathbb{Q}M \simeq \mathbb{Q}G$ as $(H_0 \times G^{op})$-bimodule.

- $C_{\mathbb{Z}G}(H_0)$ is the centralizer in $\mathbb{Z}G$ of H_0.

- $LFR(C_{\mathbb{Z}G}(H_0))$ is the set of isomorphism classes of locally free right $C_{\mathbb{Z}G}(H_0)$-ideals in $C_{\mathbb{Z}G}(H_0)$; i.e. the right ideals which become free right modules of rank one at each localization.

- $LFR_{\mathbb{Z}G}(C_{\mathbb{Z}G}(H_0))$ is the set of isomorphism classes of locally free right $C_{\mathbb{Z}G}(H_0)$-ideals $\mathcal{A}$ in $C_{\mathbb{Z}G}(H_0)$ with $\mathcal{A} \cdot \mathbb{Z}G \simeq \mathbb{Z}G$ as right $\mathbb{Z}G$-module.

In case $C_{\mathbb{Z}G}(H_0)$ is commutative, these are just the class groups with the above conditions ([Fr; 73]). Recall that for a commutative ring, the class group consists of the isomorphism classes of invertible ideals.

The above comments give the appropriate interpretations for subgroups of $U(\mathbb{Z}G)$.

Similar definitions will be used for other coefficient domains.

2.6. Lemma. *Let G be a finite group and let U_0 be a finite p–subgroup of G. The natural map*

$$Out_G(U_0) \longrightarrow Out(U_0, \mathbb{Z}G)$$

is injective.

The proof follows immediately from

2.7. Lemma. *Let R be a complete Dedekind domain with $pR \neq R$ for the rational prime p. Let U be a finite p–subgroup of the finite group G. If U is conjugate to a subgroup U_0 of G in $V(RG)$ then U is conjugate to U_0 in G.*

This is exactly the corollary to Coleman's result in section II§2.1.

Extending Fröhlich's localization sequence, we have

2.8. Lemma.

(a) *An $(H_0 \times G^{op})$–bimodule M with isomorphism class in $Picent(H_0, \mathbb{Z}G)$ is locally isomorphic to $\mathbb{Z}G$ as $H_0 \times G^{op}$ – module iff there is an $(\mathcal{A}) \in LFR(C_{\mathbb{Z}G}(H_0))$ such that $M \simeq \mathcal{A} \cdot \mathbb{Z}G$.*

(b) *An $(H_0 \times G^{op})$–bimodule M with isomorphism class in $Outcent(H_0, \mathbb{Z}G)$ is locally isomorphic to $\mathbb{Z}G$ as $H_0 \times G^{op}$–module iff there is an $(\mathcal{A}) \in LFR_{\mathbb{Z}G}(C_{\mathbb{Z}G}(H_0))$ such that $M \simeq \mathcal{A} \cdot \mathbb{Z}G$.*

Since projective $\mathbb{Z}G$-modules are locally free, we have a natural isomorphism $Picent(H_0, \hat{\mathbb{Z}}_pG) \simeq Outcent(H_0, \hat{\mathbb{Z}}_pG)$.

If $H_0 = G$, then the above results specialize to Fröhlich's localization sequence for invertible central bimodules ([Fr; 73] and §1).

PROOF: Let M be an $H_0 \times G^{op}$ – module with isomorphism class belonging to $Picent(H_0, G)$. Since $\mathbb{Q}M \simeq \mathbb{Q}G$ we may assume that $\mathbb{Q}M = \mathbb{Q}G$. Multiplying by a suitable integer we may also assume that $M \subseteq \mathbb{Z}G$.

Since there is an isomorphism $M_p \overset{\alpha_p}{\simeq} \mathbb{Z}_pG$ for all primes p, there are units $a_p \in \mathbb{Q}G$ centralizing H_0 with $M_p = a_p\mathbb{Z}_pG$ as $H_0 \times G^{op}$–bimodules, these isomorphisms α_p extend to automorphisms of $\mathbb{Q}G$ which are left multiplications if one looks only at the right structure. a_p centralizes H_0 since α_p are bimodule homomorphisms. Since $M \subseteq \mathbb{Z}G$ we see that $a_p \in \hat{\mathbb{Z}}_pG$ for all p. Observe that a_p is only defined up to right–multiplication by a unit b_p in the centralizer of H_0 in $\hat{\mathbb{Z}}_pG$ and a_p can be chosen to be 1 for almost every prime p.

We now define in the spirit of an idèlic description of an ideal (cf. section I §2)

$$\mathcal{A} := \bigcap_p (a_p C_{\hat{\mathbb{Z}}_p G}(H_0) \cap \mathbb{Q}G).$$

Since M is a full lattice in $\mathbb{Q}G$ we argue ([Re; 75, (4.22) Theorem])

$$(\mathcal{A} \cdot \mathbb{Z}G)_p = \mathcal{A}_p \cdot \hat{\mathbb{Z}}_p G = a_p C_{\hat{\mathbb{Z}}_p G}(H_0)\hat{\mathbb{Z}}_p G = a_p \hat{\mathbb{Z}}_p G.$$

Here we used the fact that $\mathcal{A}$ is locally free.

If $\mathcal{B}$ is isomorphic to $\mathcal{A}$ via ψ as $C_{\mathbb{Z}G}(H_0)$–right–module we may define an isomorphism of (H_0, G)–bimodules via:

$$\phi : \mathcal{A} \cdot \mathbb{Z}G \longrightarrow \mathcal{B} \cdot \mathbb{Z}G$$

$$a \cdot g \longrightarrow \psi(a) \cdot g.$$

This is well defined since here $\mathcal{A} \cdot \mathbb{Z}G \simeq \mathcal{A} \otimes \mathbb{Z}G$, $\mathcal{A}$ being projective. Moreover we only need to verify this locally since the multiplication is an isomorphism locally and comes from a global morphism. But locally the statement is clear.

On the other hand given an ideal $\mathcal{A}$ with isomorphism class belonging to $LFR(C_{\mathbb{Z}G}(H_0))$. Then we have

$$(\mathcal{A} \cdot \mathbb{Z}G)_p \simeq a_p C_{\hat{\mathbb{Z}}_p G}(H_0)\hat{\mathbb{Z}}_p G \simeq a_p \hat{\mathbb{Z}}_p G$$

for some $a_p \in C_{\mathbb{Z}G}(H_0)$ since $\mathcal{A}$ is locally free. We consider the $H_0 \times G^{op}$ module

$$M := \bigcap_p (a_p \hat{\mathbb{Z}}_p G \cap \mathbb{Q}G).$$

Since $M_p = a_p \hat{\mathbb{Z}}_p G$ we see that M is locally free as right G–module ([Re; 75, (4.22)]). M can be identified with $\mathcal{A} \otimes_{C_{\mathbb{Z}G}(H_0)} \mathbb{Z}G$, since $\mathcal{A}$ is locally free, and so

$$M = \mathcal{A} \cdot \mathbb{Z}G.$$

For part two we note that, if $\mathcal{A}$ has isomorphism class in the group $LFR_{\mathbb{Z}G}(C_{\mathbb{Z}G}(H_0))$, then

$$M|_G = \mathcal{A} \cdot \mathbb{Z}G|_G \simeq \mathbb{Z}G|_G$$

as right G–modules and so M is free as G–module. Conversely, if as right G–modules

$$\mathbb{Z}G \simeq M = \mathcal{A} \cdot \mathbb{Z}G,$$

then $\mathcal{A}$ has isomorphism class in $LFR_{\mathbb{Z}G}(C_{\mathbb{Z}G}(H_0))$.   □

In order to apply Lemma 2.8 to special groups, we have to recall *Milnor's Mayer-Vietoris sequence*, which is the algebraic analogue of Eilenberg-Steenrod's topological Mayer-Vietoris sequence.

Let

(42)
$$
\begin{array}{ccc}
\Lambda & \xrightarrow{\pi_1} & \Lambda_1 \\
\downarrow \pi_2 & & \downarrow \varphi_1 \\
\Lambda_2 & \xrightarrow{\varphi_2} & \Lambda'
\end{array}
$$

be a pullback diagram of rings and ring homomorphisms, and *assume* that at least one of the maps φ_i, $1 = 1, 2$ is surjective.

We recall, that the Grothendieck group $K_0(\Lambda)$ consists of the isomorphism classes of projective left Λ−modules under the direct sum with the relation $[P] = [Q]$ if $P \oplus \Lambda^{(n)} \simeq Q \oplus \Lambda^{(n)}$ for some n. The Eichler condition [Re; 75, (34.3)] (cf. below) ensures that $P \oplus \Lambda^{(n)} \simeq Q \oplus \Lambda^{(n)}$ implies $P \simeq Q$. It is a rather technical condition; however it is satisfied for integral group rings RG, if KG does not contain (2×2)−matrix rings over non commutative skewfields - more precisely skewfields, which are ramified at every infinite prime of K.

2.9. Theorem. *(Milnor [Mil; 71]): Under the above assumptions the following sequence is exact:*

$$
K_1(\Lambda) \to K_1(\Lambda_1) \oplus K_1(\Lambda_2) \to K_1(\Lambda') \to K_0(\Lambda) \to K_0(\Lambda_1) \oplus K_0(\Lambda_2) \to K_0(\Lambda'),
$$

where $K_1(\Lambda)$ is the Whitehead group of Λ and $K_0(\Lambda)$ is the Grothendieck group of projective modules.

By $K_1(\Lambda)$ we denote the Whitehead group of Λ [Re; 75][Ba; 64][C-R2; 87]:

$$
Gl(\Lambda) := lim.inj._n Gl(n, \Lambda) \text{ and } K_1(\Lambda) = Gl(\Lambda)/Gl(\Lambda)',
$$

$Gl(\Lambda)'$ being the commutator subgroup (cf. [Mil; 71])

In case of orders the result has been simplified by Reiner and Ullom [R-U; 74]:

2.10. Corollary. *Let R be a Dedekind domain with field of fractions K, and let Λ be an order in a semi simple K-algebra A, satisfying the Eichler condition. Assume further, that in the above pullback diagram (42) Λ_1 and Λ_2 are R-orders in semi simple K-algebras, Λ' is a finite ring and either φ_1 or φ_2 is surjective. We put $U^*(\Lambda_i) = \varphi_i(U(\Lambda_i))$, where $U(\Lambda_i)$ denotes the unit group of Λ_i. Then there is an exact sequence*

$$
1 \longrightarrow U^*(\Lambda_1) \cdot U^*(\Lambda_2) \longrightarrow U(\Lambda') \xrightarrow{\delta} Cl(\Lambda) \longrightarrow Cl(\Lambda_1) \oplus Cl(\Lambda_2) \longrightarrow 1.
$$

The map δ needs some explanations: Let $\delta(u) = (\Lambda_u)$ for $u \in U(\Lambda')$, then

$$
\Lambda_u := \{(\lambda_1, \lambda_2) : \lambda_i \in \Lambda_i, \varphi_1(\lambda_1) = u \cdot \varphi_2(\lambda_2)\}.
$$

Here $Cl(\Lambda)$ is the class group of locally free one sided Λ-modules $[P]$ with $KP \simeq K\Lambda$ under the addition $[P] + [Q] = [X]$ if $P \oplus Q \simeq X \oplus \Lambda$ (see [Re; 75, p. 343]).[25]

We point out, that a similar Mayer-Vietoris sequence for invertible bi-modules has been used in [R-S; 87 1], and it has been formalized in [G-R; 88].

Before we come to explicit examples let us point out some consequences.

2.11. Remark. Let P be a finite p-group, and let H be a finite subgroup of $V(\mathbb{Z}P)$; we want to classify the conjugacy classes in $V(\mathbb{Z}P)$ of such subgroups. Then H is also a finite subgroup of $V(\hat{\mathbb{Z}}_pP)$, and so by Theorem VI 1.1, H is conjugate in $V(\hat{\mathbb{Z}}_pP)$ to a subgroup H_0 of P. The bimodules $_{H_0}\hat{\mathbb{Z}}_pP_P$ and $_H\hat{\mathbb{Z}}_pP_P$ thus differ by an automorphism of H_0. Now two bimodules M_p and N_p for $H_0 \times G^{op}$ over the local ring $\mathbb{Z}_p$ are isomorphic iff the corresponding p-adic completions are isomorphic. Consequently — up to automorphisms of H_0 – the conjugacy classes of finite subgroups H of $V(\mathbb{Z}P)$ are parameterized by $Outcent(H_0, \mathbb{Z}P)$.

Summarizing this discussion, we have

2.12. Lemma. *Let P be a p-group, then the conjugacy classes of finite subgroups H of $V(\mathbb{Z}P)$ isomorphic to a finite subgroup H_0 of P are parameterized — modulo automorphisms of P and of H_0 — by $LFR_{\mathbb{Z}P}(C_{\mathbb{Z}P}(H_0))$.*

Whenever the *Zassenhaus conjecture* (section IX) holds for a class of subgroups of the finite group G, a similar statement can be made.

2.13. Claim. *If $|G|$ is invertible in the local Dedekind domain S the group $Picent(H_0, SG)$ is trivial.*

PROOF: Let M be an $H \times G^{op}$ – lattice such that the module isomorphism class (M) is an element in $Picent(H_0, RG)$. If K is the quotient field of S we have $KM \simeq KG$ as $H \times G^{op}$ modules. But since the order of G is invertible in S, $S(H \times G^{op})$–lattices M and N are isomorphic over $S(H \times G^{op})$ if and only if KM and KN are isomorphic over $K(H \times G^{op})$.   □

In paragraph 3 we shall apply this to various examples.

§3   Conjugacy of Finite Subgroups to Group Bases

Since for a p–group G, the integral cohomology is determined p–adically (cf. VIII.11), where its variety coincides with the variety of the p-adic cohomology of $V(\hat{\mathbb{Z}}_pG)$ (cf. II. §3), is it possible, that for p-groups the varieties of $H^*(G, \mathbb{F}_p)$ and of $H^*(V(\mathbb{Z}G), \mathbb{F}_p)$ are also related? It should be noted though, that there is no obvious connection between $V(\mathbb{Z}G)$ and $V(\hat{\mathbb{Z}}_pG)$. By a result of Quillen [Qui; 71], the irreducible components of the variety of

---

25   Note that $[P] = [Q]$ in $Cl(\Lambda)$ iff $P \oplus \Lambda^n \simeq Q \oplus \Lambda^n$.

$H^*(V(\mathbb{Z}G), \mathbb{F}_p)$ are in bijection to the conjugacy classes of maximal elementary abelian p-subgroups of $V(\mathbb{Z}G)$.

The above question relates to:

For G a p-group, how are the conjugacy classes of maximal elementary abelian p-subgroups of $V(\mathbb{Z}G)$ related to those of G?

We shall elaborate on this in the next section; however, we state here an example, which shows, that the cohomology varieties of the p-group G and that of the unit group of its integral group ring $\mathbb{Z}G$ are not as intimately related as they are p-adically.

3.1. Lemma. *The maximal elementary abelian 2-subgroups of the dihedral group D of order 8 are all Klein's four groups, and there are two conjugacy classes. In $V(\mathbb{Z}D)$ there are three conjugacy classes of Klein's four groups. However, in $V(\hat{\mathbb{Z}}_2 D)$ there are only two conjugacy classes of Klein's four groups.*

The proof will be postponed until paragraph 3.4.

3.2. The Symmetric Group of Order 6

3.3. Lemma. *In $V(\mathbb{Z}S_3)$ there is an involution ι which is not conjugate in $V(\hat{\mathbb{Z}}_2 S_3)$ to a group element in S_3. All units of order 3 in $\mathbb{Z}S_3$ are conjugate to some group element. Moreover ι is not part of a group basis.*

According to Hughes and Pearson [H-P; 72] the element in question is

$$\iota = (12) + 3(13) - 3(23) - 3(123) + 3(132).$$

(Here the group elements are written in their cycle decomposition.)

We shall however give a theoretical argument, which shows, that such an involution exists.

PROOF: The group ring is a pullback

$$
\begin{array}{ccccccccc}
0 & \longrightarrow & I(<(123)>)S_3 & \longrightarrow & \mathbb{Z}S_3 & \xrightarrow{\pi_1} & \mathbb{Z}C_2 & \longrightarrow & 0 \\
 & & \| & & \downarrow \pi_2 & & \downarrow \phi_1 & & \\
0 & \longrightarrow & I(<(123)>)S_3 & \longrightarrow & \Lambda_2 & \xrightarrow{\phi_2} & \mathbb{F}_3 C_2 & \longrightarrow & 0
\end{array}
$$

as one gets from the decomposition of the rational group algebra into two one–dimensional and one two–dimensional representations (cf.[Ro; 72 1]). Here $\Lambda_2 = \begin{pmatrix} \mathbb{Z} & \mathbb{Z} \\ 3\mathbb{Z} & \mathbb{Z} \end{pmatrix}$ and ϕ_2 is reduction modulo $\begin{pmatrix} 3\mathbb{Z} & \mathbb{Z} \\ 3\mathbb{Z} & 3\mathbb{Z} \end{pmatrix}$. The unit group of $\mathbb{F}_3 C_2$ is Klein's four-group and all these units come from units of $\mathbb{Z}C_2$. Consequently, a finite subgroup U of $V(\mathbb{Z}S_3)$ is conjugate to a subgroup of S_3 iff it is locally conjugate to a subgroup of S_3.

At the prime 3 we have $O_{3'}(S_3) = 1$, and hence all group bases are conjugate to S_3 (cf. section VI Theorem VI.1, [Sc; 87, Sc; 90]).

Let η be an element of order 3 in $V(\mathbb{Z}S_3)$, then it must project onto an element of order 3 in Λ_2. Thus it is conjugate in $M_2(\mathbb{Z})$ to a 3-cycle in Σ_3, since there is only one 2-dimensional faithful $\mathbb{Z}$-representation M of the cyclic group of order 3, C_3. We still have to show that η must be conjugate in Λ_2 to a 3-cycle in Σ_3. For this we have to analyze the two-dimensional representation of C_3 more carefully: Since $M = (\mathbb{Z} \times \mathbb{Z})^{tr}$ is a Λ_2 – and also a $M_2(\mathbb{Z})$-lattice, it has at the prime 3 a unique maximal C_3-sublattice $M_0 = (\hat{\mathbb{Z}}_3 \times 3\hat{\mathbb{Z}}_3)^{tr}$, which is left invariant under any C_3-automorphism of M. Thus the conjugating element lies in Λ_2.

For the prime 2 we see that the group ring looks like

$$\hat{\mathbb{Z}}_2 S_3 = \{(a_1, a_1 + 2a_2, \begin{pmatrix} a_3 & a_4 \\ a_5 & a_6 \end{pmatrix}) \mid a_i \in \hat{\mathbb{Z}}_2 \ \forall i = 1, .., 6\},$$

and we have an involution

$$\iota = (1, -1, \begin{pmatrix} 1 & 0 \\ 0 & -1 \end{pmatrix}))$$

that is not conjugate in $\hat{\mathbb{Z}}_2 S_3$ to an involution in a group basis, since it is a decomposable representation of C_2 and the representation of S_3 yields an indecomposable representation for its involutions like

$$e = (1, -1, \begin{pmatrix} 0 & 1 \\ 1 & 0 \end{pmatrix})).$$

Because of the above remark a unit of finite order is determined, if we describe it locally. We now observe that in $\mathbb{Q}S_3$ the elements ι and e are conjugate. Thus in $V(\mathbb{Z}S_3)$ there is an involution u which is e at the primes $p \neq 2$ and ι at 2. Then u and ι are rationally conjugate by an element a. However u cannot be part of any group bases, and so conjugation by a cannot normalize $\mathbb{Z}G$. Moreover $_{<u>}|\mathbb{Z}S_3$ is not free since at 2 the element u is ι, which does not act freely.

$\square$

3.4. The Dihedral Group of Order 8

Let $D = <a, b|a^4 = b^2 = baba = 1>$ be the dihedral group of order 8 and $V = <\bar{a}, \bar{b}>$ be the quotient modulo the center, Klein's 4 group.

3.5. Lemma. *In the augmented units of $\mathbb{Z}D$ there are two conjugacy classes of group bases. Every finite subgroup of the augmented units $V(\mathbb{Z}D_8)$ is part of a group basis.*

Once again we present the group ring as a pullback

$$\begin{array}{ccccccccc}
0 & \longrightarrow & I(<a^2>)D & \longrightarrow & \mathbb{Z}D & \overset{\sigma}{\longrightarrow} & \mathbb{Z}V & \longrightarrow & 0 \\
& & \| & & \downarrow \tau & & \downarrow \psi & & \\
0 & \longrightarrow & I(<a^2>)D & \longrightarrow & \Lambda & \overset{\phi}{\longrightarrow} & \mathbb{F}_2V & \longrightarrow & 0, \text{26}
\end{array}$$

with σ the map induced from the natural projection of D to V, Klein's four-group. Λ is the subring of $\begin{pmatrix} \mathbb{Z} & \mathbb{Z} \\ 2\mathbb{Z} & \mathbb{Z} \end{pmatrix}$ consisting of matrices with entries in the main diagonal differing by integral multiples of 2. There we have a representation by

$$a \longrightarrow \begin{pmatrix} -1 & -1 \\ 2 & 1 \end{pmatrix} \text{ and } b \longrightarrow \begin{pmatrix} -1 & 0 \\ 2 & 1 \end{pmatrix}.$$

We want to use our interpretation as bimodules and the generalization of Fröhlich's localization sequence. In practice we use Milnor's Mayer-Vietoris sequence. Therefore certain class–groups should be calculated.

3.6. Claim. $Cl(\mathbb{Z}V) = 1$.

PROOF: We write a pullback diagram as follows:

$$\begin{array}{ccc} \mathbb{Z}V & \longrightarrow & \mathbb{Z} < b > \\ \downarrow & & \downarrow \\ \mathbb{Z} < b >_0 & \longrightarrow & \mathbb{F}_2 < b > \end{array},$$

where $\mathbb{Z} < b >= \mathbb{Z} < b >_0$ except that a acts as -1 on $\mathbb{Z} < b >_0$ and it acts as 1 on $\mathbb{Z} < b >$. Obviously $U(\mathbb{F}_2 < b >)$ is of order 2 and hence is the image of $U(\mathbb{Z} < b >)$. By Rim's Theorem ([C-R2; 87, Theorem 50.2]), which is a consequence of Milnor's Mayer Vietoris sequence, we have $Cl(\mathbb{Z} < b >) = 1$ and therefore $Cl(\mathbb{Z}V) = 1$ by Milnor's Mayer Vietoris sequence. $\square$

3.7. Claim. *If we write $C_X(Y)$ for the centralizer of Y in X and $Z(R)$ for the center of R we get*

$$|Cl(C_{\mathbb{Z}D}(b))| = |Cl(Z(\mathbb{Z}D))| = 2.$$

PROOF: The centralizer of a subgroup H of G in $\mathbb{Z}G$ is the R–linear span of the H–conjugacy class sums of elements in G as one sees by elementary calculations.

We set $C := C_{\mathbb{Z}D}(b)$ and $Z = Z(\mathbb{Z}D)$. Then

$$C = < 1, a^2, a + a^3, (a + a^3)b, b, a^2b >_{\mathbb{Z}};$$

$$Z = < 1, a^2, (a + a^3), (a + a^3)b, (1 + a^2)b >_{\mathbb{Z}}.$$

Under $\sigma : D \longrightarrow V_4$ the elements a and a^3 are identified and we get

$$\sigma(C) = \mathbb{Z} < b > + 2\mathbb{Z}V \text{ and } \sigma(Z) = \mathbb{Z} + 2\mathbb{Z}V.$$

So we get pullbacks

$$\begin{array}{ccccccc} C & \longrightarrow & \sigma(C) & & Z & \longrightarrow & \sigma(Z) \\ \downarrow & & \downarrow & \text{and} & \downarrow & & \downarrow \\ \tau(C) & \longrightarrow & \mathbb{F}_2 < b > & & \tau(Z) & \longrightarrow & \mathbb{F}_2 \end{array}.$$

The unit groups here are easy to determine and they all come from units in $\sigma(C)$ and $\sigma(Z)$ respectively.

The Mayer–Vietoris sequence tells us now that

$$Cl(C) = Cl(\sigma(C)) \oplus Cl(\tau(C)); \; Cl(Z) = Cl(\sigma(Z)) \oplus Cl(\tau(Z)).$$

For this we observe that the idempotents η_0 and η_1 defining σ and τ lie in C and Z, since they are central. Since Λ is a full sublattice of the rational 2×2 matrix ring, $\tau(Z) = \mathbb{Z}$ belongs to the center of this lattice. Similarly $\tau(C) \simeq \mathbb{Z} < b >$ (the images of b and 1 are linearly independent, and so one gets a monomorphism induced from τ onto Λ). Using again Rim's Theorem and Milnor's Mayer–Vietoris sequence we can omit the τ contribution.

By the presentation of V as a pullback we have pullbacks

$$
\begin{array}{ccc}
2\mathbb{Z}V & \longrightarrow & 2\mathbb{Z} < b > \\
\downarrow & & \downarrow \\
2\mathbb{Z} < b >_0 & \longrightarrow & 2\mathbb{Z}/(4\mathbb{Z} < b >)
\end{array}
$$

and resulting pullbacks

$$
\begin{array}{ccc}
\mathbb{Z} < b > +2\mathbb{Z}V & \longrightarrow & \mathbb{Z} < b > \\
\downarrow & & \downarrow \\
\mathbb{Z} < b >_0 & \longrightarrow & \mathbb{Z}/(4\mathbb{Z} < b >)
\end{array}
$$

and

$$
\begin{array}{ccc}
\mathbb{Z} + 2\mathbb{Z}V & \longrightarrow & \mathbb{Z} + 2\mathbb{Z} < b > \\
\downarrow & & \downarrow \\
(\mathbb{Z} + 2\mathbb{Z} < b >)_0 & \longrightarrow & \mathbb{Z}/4\mathbb{Z} + 2\mathbb{Z}/4\mathbb{Z} < b >
\end{array} ;
$$

a acts as -1 on the modules assigned by the index 0.

Since $2\mathbb{Z} < b >$ is a pullback

$$
\begin{array}{ccc}
2\mathbb{Z} < b > & \longrightarrow & 2\mathbb{Z} \\
\downarrow & & \downarrow \\
(2\mathbb{Z})_0 & \longrightarrow & 2\mathbb{Z}/4\mathbb{Z}
\end{array} ,
$$

we get the pullback

$$
\begin{array}{ccc}
\mathbb{Z} + 2\mathbb{Z} < b > & \longrightarrow & \mathbb{Z} \\
\downarrow & & \downarrow \\
(\mathbb{Z})_0 & \longrightarrow & \mathbb{Z}/4\mathbb{Z}
\end{array} ,
$$

and looking at the units in $\mathbb{Z}$ we see that

$$Cl(\mathbb{Z} + 2\mathbb{Z} < b >) = 1,$$

if we use again Milnor's Mayer Vietoris sequence.

We are now able to attack the centralizer of b. The units in $\mathbb{Z}/4\mathbb{Z} < b >$ are $\{1, 3, b, 3b, 2+b, 2+3b, 1+2b, 3+2b\}$ (all of them are involutions). The units in $\mathbb{Z} < b >$ and $\mathbb{Z} < b >_0$ are $\{1, -1, b, -b\}$.

The Mayer Vietoris sequence appears now as

$$1 \longrightarrow C_2^{(2)} \longrightarrow C_2^{(3)} \longrightarrow Cl(C_{\mathbb{Z}D}(b)) \longrightarrow 1 \oplus 1 \longrightarrow 1$$

and therefore $|Cl(C_{\mathbb{Z}D}(b)| = 2$.

For the center of the group ring we observe that the only units in $\mathbb{Z} < b >$ are $\{1, -1, b, -b\}$ and therefore $U(\mathbb{Z} + 2\mathbb{Z} < b >) = \{1, -1\}$. Comparing it with above, we see that $U((\mathbb{Z} + 2\mathbb{Z} < b >)/4\mathbb{Z}) = \{1, 3, 1+2b, 3+2b\}$, and so the Mayer Vietoris sequence yields the exact sequence

$$1 \longrightarrow C_2 \longrightarrow C_2 \times C_2 \longrightarrow Cl(Z(\mathbb{Z}D)) \longrightarrow 1 \oplus 1 \longrightarrow 1,$$

and we obtain the class number of $Z(\mathbb{Z}D)$, it is again of order 2.

For the other involutions we obtain:

$$C_{\mathbb{Z}D}(ab) = < 1, a + a^3, a^2, (a + a^3)b, a^2b, b >_{\mathbb{Z}} = C; \quad Z(\mathbb{Z}D) = C_{\mathbb{Z}D}(a^2).$$

q.e.d. □

So we see that there are exactly two stable isomorphism classes of ideals of $Z(\mathbb{Z}D)$ and of $C_{\mathbb{Z}D}(b)$. This proves the lemma.

VIII Locally isomorphic group rings

In [Da; 64,1] E. C. Dade has constructed two non isomorphic metabelian groups G and H of order $p^3 \cdot q^6$ for suitable prime numbers p and q, which have isomorphic group rings over *every* field. An analysis of his proof shows that also the group rings over $\hat{\mathbb{Z}}_p$ are isomorphic for these pairs of groups for all primes p.

The aim of this section is twofold:

(a) We construct two non isomorphic metabelian groups G and H of order $5 \cdot 19^2 \cdot 29^2$, which have isomorphic group rings over $\hat{\mathbb{Z}}_p$ for every prime p.

(b) We construct two non isomorphic groups G and H which have isomorphic group rings over R, where R is a large enough extension of $\hat{\mathbb{Z}}_p$; however, the group rings over $\hat{\mathbb{Z}}_p$ for a suitable prime p are not isomorphic.

In order to construct the groups, we note that the cyclic group C_5 of order 5 has two non isomorphic two dimensional $\mathbb{F}_p C_5$–modules M_p and N_p, provided 5 divides $p + 1$. The smallest primes for which this happens are $p = 19$ and $p = 29$. We then form the semi direct products

$$G = M_{19} \oplus M_{29} \rtimes C_5$$

and

$$H = M_{19} \oplus N_{29} \rtimes C_5.$$

VIII.1. Note. These groups are non isomorphic [Asch; 87, 10.3].

We shall show, that they have isomorphic group rings

$$\hat{\mathbb{Z}}_p G \simeq \hat{\mathbb{Z}}_p H$$

for every prime p.

More precisely, let

$$A_{19} = < a, b, c \mid a^{19}, b^{19}, c^5, [a, b], \ {}^c a = b^9, \ {}^c b = a^2 \cdot b^4 > \text{ and}$$

$$A_{29} = < \alpha, \beta, \gamma \mid \alpha^{29}, \beta^{29}, \gamma^5, [\alpha, \beta], \ {}^\gamma \alpha = \beta^{14}, \ {}^\gamma \beta = \alpha^2 \cdot \beta^5 > .$$

Then A_{19} and A_{29} can be viewed as subgroups of the affine groups

$$\mathbb{F}_{19^2} \rtimes \mathbb{F}^*_{19^2} \text{ and } \mathbb{F}_{29^2} \rtimes \mathbb{F}^*_{29^2}.$$

and the matrices $M = \begin{pmatrix} 0 & 2 \\ 9 & 4 \end{pmatrix}$, $N = \begin{pmatrix} 0 & 2 \\ 14 & 5 \end{pmatrix}$ act as a generator c and γ resp. of C_5, which we fix. Then A_{19} has order $19 \cdot 19 \cdot 5$ and A_{29} has order $29 \cdot 29 \cdot 5$.

VIII.2. Claim. *For the group rings over Q we have:*

$$Q\,A_{19} \simeq Q \prod Q(\zeta_5) \prod_1^4 (Q(\zeta_{19}))_5,$$

and

$$Q\,A_{29} \simeq Q \prod Q(\zeta_5) \prod_1^6 (Q(\zeta_{29}))_5,$$

where ζ_i is a primitive i-th root of unity.

PROOF: A_{19} has 20 cyclic subgroups of order 19 generated by $a^i b^j$, which fall into four classes under the operation of $< c >$. Altogether A_{19} has 6 conjugacy classes of cyclic subgroups, and hence 6 non isomorphic irreducible Q-representations. Similarly, A_{29} has 8 conjugacy classes of cyclic subgroups.

We only treat A_{19} in detail. The group ring $Q\,A_{19}$ decomposes as

$$(Q\,I(< a, b >){\cdot}< c >) \times Q\,C_5,$$

where $I(< a, b >)$ is the augmentation ideal of $< a,\ b >$. The first factor J decomposes into 4 algebraically conjugate representations, one of which we give explicitly: $\zeta := \zeta_{19}$ is a primitive 19-th root of unity. Then one of the 5-dimensional representations of A_{19} over $I\!\!F^2$ is given by

$$a \longrightarrow \begin{pmatrix} \zeta^{-1} & 0 & 0 & 0 & 0 \\ 0 & \zeta^{-4} & 0 & 0 & 0 \\ 0 & 0 & \zeta^4 & 0 & 0 \\ 0 & 0 & 0 & \zeta^1 & 0 \\ 0 & 0 & 0 & 0 & 1 \end{pmatrix}, b \longrightarrow \begin{pmatrix} \zeta^8 & 0 & 0 & 0 & 0 \\ 0 & \zeta^{-8} & 0 & 0 & 0 \\ 0 & 0 & \zeta^{-2} & 0 & 0 \\ 0 & 0 & 0 & 1 & 0 \\ 0 & 0 & 0 & 0 & \zeta^2 \end{pmatrix},$$

$$c \longrightarrow \begin{pmatrix} 0 & 1 & 0 & 0 & 0 \\ 0 & 0 & 1 & 0 & 0 \\ 0 & 0 & 0 & 1 & 0 \\ 0 & 0 & 0 & 0 & 1 \\ 1 & 0 & 0 & 0 & 0 \end{pmatrix}.$$

The action of c on a is $c \cdot a \cdot c^{-1}$.

This shows that $Q\,A_{19}$ has the form which we have claimed. □

We can now define the groups G and H :

$$G =< a, b, \alpha, \beta, c|\quad a^{19}, b^{19}, [a, b], \alpha^{29}, \beta^{29}, [\alpha, \beta], c^5,$$
$${}^c a = b^9, \ {}^c b = a^2 \cdot b^4, \ {}^c \alpha = \beta^{14}, \ {}^c \beta = \alpha^2 \cdot \beta^5 >,$$

$$H =< a, b, \alpha, \beta, c|\quad a^{19}, b^{19}, [a, b], \alpha^{29}, \beta^{29}, [\alpha, \beta], c^5,$$
$${}^c a = b^9, \ {}^c b = a^2 \cdot b^4, \ {}^c \alpha = \alpha^{28} \cdot \beta^{12}, \ {}^c \beta = \alpha^{10} \cdot \beta^{24} > .$$

(We have used the same name for the generators in G and H. If there is a possibility for confusion we shall distinguish them by a subscript "G" und "H" resp.) G and H have both order $5 \cdot 19 \cdot 19 \cdot 29 \cdot 29$. It should be noted that the difference between G and H is only the action of c on α and β : it is the action of c^2 on G.

The above arguments have shown: The groups G and H are not isomorphic (VIII.1).

In order to show that the group rings $\hat{\mathbb{Z}}_p G$ and $\hat{\mathbb{Z}}_p H$ are isomorphic, we first consider the rational group algebras.

VIII.3. Claim. *G and H are pullbacks*

$$
\begin{array}{ccc}
A_{19} & \longrightarrow & C_5 \\
\uparrow & & \uparrow \\
G & \longrightarrow & A_{29}
\end{array}
\quad and \quad
\begin{array}{ccc}
A_{19} & \longrightarrow & C_5 \\
\uparrow & & \uparrow \\
H & \longrightarrow & A_{29}
\end{array} .
$$

The group ring of G over $\mathbb{Z}$ thus has a factor Γ, which is the induced pullback

$$
\begin{array}{ccc}
\mathbb{Z}A_{19} & \longrightarrow & \mathbb{Z}C_5 \\
\uparrow & & \uparrow \\
\Gamma_G & \longrightarrow & \mathbb{Z}A_{29}
\end{array}
$$

The whole group ring of G over $\mathbb{Z}$ is thus given as

$$
(43) \quad
\begin{array}{ccccccccc}
0 & \longrightarrow & I(A_{19}) \cdot G \cap I(A_{29}) \cdot G & \longrightarrow & \mathbb{Z}G & \longrightarrow & \Gamma_G & \longrightarrow & 0 \\
& & \| & & \downarrow & & \downarrow & & \\
0 & \longrightarrow & I(A_{19}) \cdot G \cap I(A_{29}) \cdot G & \longrightarrow & \Lambda_G & \longrightarrow & \Delta_G & \longrightarrow & 0
\end{array}
$$

Here Λ_G is the projection onto the component, where neither A_{19} nor A_{29} does act trivially. Then

$$
(44) \qquad \mathbb{Q} \otimes \Lambda_G = \prod_{1}^{24} (\mathbb{Q}[\zeta_{19 \cdot 29}])_5 .
$$

The group rings $\mathbb{Q}G$ and $\mathbb{Q}H$ are isomorphic.

PROOF: The description of $\mathbb{Z}G$ follows from (38). The structure of $\mathbb{Q} \otimes \Lambda_G$ is an easy application of Clifford theory (cf. Chapter XIII), using Claim VIII.2. In order to see that the rational group algebras are isomorphic, we note that $\mathbb{Q} \otimes \Gamma_G \simeq \mathbb{Q} \otimes \Gamma_H$ by Claim VIII.2 and surely $\mathbb{Q} \otimes \Lambda_G \simeq \mathbb{Q} \otimes \Lambda_H$. $\square$

We now turn to the integral group rings:

VIII.4. Theorem. *For every rational prime p, we have an isomorphism of the group rings*

$$
\hat{\mathbb{Z}}_p G \simeq \hat{\mathbb{Z}}_p H .
$$

PROOF: We shall treat the group rings a prime at a time:

VIII.5. Case 1. $(p, |G|) = 1$.

Then $\hat{\mathbb{Z}}_p G \simeq \hat{\mathbb{Z}}_p H$ since the rational group algebras are isomorphic.

VIII.6. Case 2. $p = 5$.

In this case, the exact sequence in (43) splits; i.e. $\hat{\mathbb{Z}}_p G \simeq \Lambda_G \prod \Gamma_G$. Moreover, Λ_G is a sum of blocks of defect zero. In fact even the exact sequence

$$0 \longrightarrow I_{\hat{\mathbb{Z}}_5}(A_{19} \times A_{29}) \cdot G \longrightarrow \hat{\mathbb{Z}}_5 G \longrightarrow \hat{\mathbb{Z}}_5 C_5 \longrightarrow 0$$

is two-sided split, and $I_{\hat{\mathbb{Z}}_5}(A_{19} \times A_{29}) \cdot G$ is a sum of blocks of defect zero. In order to see this, we pass to an integrally closed finite extension R of $\hat{\mathbb{Z}}_5$, which contains a primitive $(19 \cdot 29)^{th}$ root of unity. If M is an irreducible non trivial $A_{19} \times A_{29}$–module, then the induced module $M\uparrow^G$ is projective of rank 5 and is itself irreducible (VIII.2). Thus it lies in a block of defect zero. Hence the group ring in (43) has the form

$$\hat{\mathbb{Z}}_5 G \simeq \prod_1^4 (\hat{\mathbb{Z}}_5[\zeta_{19}])_5 \prod_1^6 (\hat{\mathbb{Z}}_5[\zeta_{29}])_5 \prod_1^{24} (\hat{\mathbb{Z}}_5[\zeta_{19\cdot29}])_5 \prod \hat{\mathbb{Z}}_5 C_5.$$

It is now obvious, that the above factors are isomorphic for G and H. Hence

$$\hat{\mathbb{Z}}_5 G \simeq \hat{\mathbb{Z}}_5 H.$$

VIII.7. Case 3. $p = 19$ and $p = 29$.

We note that the groups G/A_{29} and H/A_{29} are *equal*, and the groups G/A_{19} and H/A_{19} are *isomorphic*.

VIII.8. Case 3,19. Let $p = 19$.

Then we have the two exact sequences:

$$0 \longrightarrow I_{\hat{\mathbb{Z}}_{19}}(A_{29}) \cdot G \longrightarrow \hat{\mathbb{Z}}_{19} G \longrightarrow \hat{\mathbb{Z}}_{19} G/A_{29} \longrightarrow 0$$
$$0 \longrightarrow I_{\hat{\mathbb{Z}}_{19}}(A_{29}) \cdot H \longrightarrow \hat{\mathbb{Z}}_{19} H \longrightarrow \hat{\mathbb{Z}}_{19} H/A_{29} \longrightarrow 0,$$

which are two sided split. Since $G/A_{29} = H/A_{29}$, we only need to show that

$$I_{\hat{\mathbb{Z}}_{19}}(A_{29}) \cdot G \simeq I_{\hat{\mathbb{Z}}_{19}}(A_{29}) \cdot H.$$

However, let M be a non-trivial irreducible $\hat{\mathbb{Z}}_{19} A_{29}$-module, then the inertia group of M in G is $A_{19} \times A_{29}$, which is the same as the inertia group of M as $\hat{\mathbb{Z}}_{19}[\zeta_{29}]$–module. Moreover it is the same for H. The results in section XIII now show, that

$$I_{\hat{\mathbb{Z}}_{19}}(A_{29}) \cdot G \simeq I_{\hat{\mathbb{Z}}_{19}}(A_{29}) \cdot H,$$

note that the rational group algebras are isomorphic.

VIII.9. Case 3,29. If $p = 29$,

then the arguments are similar, noting that in this case $G/A_{19} \simeq H/A_{19}$.

This completes the example. $\qquad\square$

VIII.10. Note. This example also provides us with two non isomorphic finite metabelian groups G and H, for which the cohomology rings $H^*(G, \mathbb{Z})$ and $H^*(H, \mathbb{Z})$ are isomorphic. This follows from the next lemma.

VIII.11. Lemma. *Let G and H be finite groups, which have locally isomorphic integral group rings; i.e. for every rational prime p, the group rings $\hat{\mathbb{Z}}_p G$ and $\hat{\mathbb{Z}}_p H$ are isomorphic; it is even enough that for every prime p the principal blocks $B_{0,G,p}$ and $B_{0,H,p}$ of $\hat{\mathbb{Z}}_p G$ and $\hat{\mathbb{Z}}_p H$ are isomorphic. Then*

$$H^*(G, \mathbb{Z}) \simeq H^*(H, \mathbb{Z}).$$

PROOF: We note that for any finite group G, $H^0(G, \mathbb{Z}) \simeq \mathbb{Z}$, and for $i \geq 1$ we have

$$H^i(G, \mathbb{Z}) \cong Ext^i_{\mathbb{Z}G}(\mathbb{Z}, \mathbb{Z}) \cong \oplus_p Ext^i_{\hat{\mathbb{Z}}_p G}(\hat{\mathbb{Z}}_p, \hat{\mathbb{Z}}_p) \cong \oplus_p Ext^i_{B_{o,G,p}}(\hat{\mathbb{Z}}_p, \hat{\mathbb{Z}}_p),$$

where the sum has to be taken only over the primes p which divide $|G|$, and hence the local isomorphism of the group ring and principal blocks resp. implies that the additive structures of $H^*(G, \mathbb{Z})$ and $H^*(H, \mathbb{Z})$ are isomorphic. As for the multiplicative structure (II §3), we note that the cup product of $H^i_{\hat{\mathbb{Z}}_p}(G, \hat{\mathbb{Z}}_p)$ with $H^i_{\hat{\mathbb{Z}}_q}(G, \hat{\mathbb{Z}}_q)$ is zero, if p and q are relatively prime. Thus the multiplicative structure of $H^*(G, \mathbb{Z})$ is determined locally.   □

I personally do not know of any two finite p-groups, which have isomorphic cohomology rings.

VIII.12. Remark. We shall next construct two non isomorphic finite metabelian groups G and H, which have isomorphic group rings $R_\wp G \simeq R_\wp H$ for every completion of the ring of algebraic integers R, which is large enough; however, the group rings $\hat{\mathbb{Z}}_p G$ and $\hat{\mathbb{Z}}_p H$ are not isomorphic for all primes p. The example is similar to the one above, however, the groups have a very simple structure:

Let

$$
\begin{aligned}
(45) \quad G &= \; < a, b, c : a^7 = b^{13} = c^3 = [a, b] = 1, \; {}^c a = a^2, \; {}^c b = b^3 >, \\
H &= \; < a, b, c : a^7 = b^{13} = c^3 = [a, b] = 1, \; {}^c a = a^4, \; {}^c b = b^3 > .
\end{aligned}
$$

$$A = <a>, B = <b> \text{ and } C = <c> .$$

We have used the same letters for the subgroups of H and G. If there is a possibility for confusion, we distinguish them by a subscript "G" and "H" resp.

VIII.13. Note. Then it is easily seen, that G and H are not isomorphic [Asch; 87, 10.3].

VIII.14. Lemma. *Let $R = \mathbb{Z}[\zeta_{7\cdot13}]$. Then for every prime ideal $\wp$ of R, the group rings $R_\wp G$ and $R_\wp H$ are isomorphic; however, the group rings $\hat{\mathbb{Z}}_p G$ and $\hat{\mathbb{Z}}_p H$ are not isomorphic.*

VIII.15. Remark. As one would expect, this shows, that for group rings a Noether-Deuring type theorem is not valid.

PROOF OF LEMMA VIII.14: We observe, that

$$(46) \qquad G/B = H/B \text{ and } G/A \simeq H/A,$$

the isomorphism is given by forming the pullback along $c \to c^{-1}$.

Let S be any commutative ring. We have a pullback diagram of groups

$$(47) \qquad \begin{array}{ccc} G/A & \longrightarrow & C \\ \uparrow & & \uparrow \\ G & \longrightarrow & G/B \end{array},$$

which induces a pullback diagram of group rings

$$(48) \qquad \begin{array}{ccc} SG/A & \longrightarrow & SC \\ \uparrow & & \uparrow \\ \Gamma_G & \longrightarrow & SG/B \end{array}.$$

Moreover, we have exact sequences

$$(49) \qquad \begin{array}{ccccccccc} 0 & \to & I_S(A)\cdot G \cap I_S(B)\cdot G & \to & SG & \to & \Gamma_G & \to & 0 \\ & & \| & & \downarrow & & \downarrow & & \\ 0 & \to & I_S(A)\cdot G \cap I_S(B)\cdot G & \to & \Lambda_G & \to & \tilde{\Lambda} & \to & 0 \end{array}.$$

(cf. [R-S; 87 3])

(a) The group algebras $\mathbb{Q}G$ and $\mathbb{Q}H$ are not isomorphic.
Let γ be the element of order 3 in $Gal(\mathbb{Q}[\zeta_7]/\mathbb{Q})$ and $Gal(\mathbb{Q}[\zeta_{13}]/\mathbb{Q})$ resp., where ζ_i is a primitive i^{th} root of unity, and the action is given as

$$^\gamma\zeta_7 = \zeta_7^2 \text{ and } ^\gamma\zeta_{13} = \zeta_{13}^3 \text{ resp.}$$

We put $K = \mathbb{Q}[\zeta_7 \cdot \zeta_{13}]^{(\gamma,\gamma)}$ to be the fixed field of $\mathbb{Q}[\zeta_{7\cdot13}]$ under the diagonal (γ,γ) in $Gal(\mathbb{Q}[\zeta_{7\cdot13}])$ and $L = \mathbb{Q}[\zeta_7\cdot\zeta_{13}]^{(\gamma,\gamma^{-1})}$. Then K and L are not isomorphic, since $\mathbb{Q}[\zeta_{7\cdot13}]$ is a Galois extension of $\mathbb{Q}$ with abelian Galois group.
However, $\mathbb{Q} \otimes \Lambda_G = (K)_3$ and $\mathbb{Q} \otimes \Lambda_H = (L)_3$. In fact, c_G and c_H act on an irreducible faithful $\mathbb{Q}(A \times B)$–module $M = \mathbb{Q}[\zeta_{7\cdot13}]$ as (γ,γ) and (γ,γ^{-1}) resp. The fixed fields of these actions are then K and L resp.

(b) The group rings $\mathbb{Q}_7 G$ and $\mathbb{Q}_7 H$ are not isomorphic.

To see this, we consider the rings

$$\mathbb{Q}_7 \otimes \Lambda_G = (\mathbb{Q}_7 \otimes K) \text{ and } \mathbb{Q}_7 \otimes \Lambda_H = (\mathbb{Q}_7 \otimes L).$$

An easy calculation shows, that $\mathbb{Q}_7[\zeta_{13}]$ is an unramified extension of $\mathbb{Q}_7$ with residue class degree 12. Now the same arguments as in the first part show, that

$$\mathbb{Q}_7 \otimes \Lambda_G = (\mathbb{Q}_7 \otimes K) \text{ and } \mathbb{Q}_7 \otimes \Lambda_H = (\mathbb{Q}_7 \otimes L)$$

and hence $\mathbb{Q}_7 G$ and $\mathbb{Q}_7 H$ can not be isomorphic.

Below we shall use Clifford theory. So let us briefly look at the situation here for $\hat{\mathbb{Z}}_7 G$: If $M = \hat{\mathbb{Z}}_7[\zeta_{13}]$ is a faithful irreducible $\hat{\mathbb{Z}}_7 B$−module, then:

VIII.16. Note. M is as $\hat{\mathbb{Z}}_7 B$−module isomorphic to its C−conjugate, and so $< c >$ lies in the inertia group of M as $\hat{\mathbb{Z}}_7 B$−modules; however, as $\hat{\mathbb{Z}}_7[\zeta_{13}]B$−module, M is not isomorphic to its C−conjugate, and so $< c >$ does not lie in the inertia group of M as $\hat{\mathbb{Z}}_7[\zeta_{13}]B$−module.

We now turn to the group rings over $R = \mathbb{Z}[\zeta_{7 \cdot 13}]$. Let $\mathbb{F} = \mathbb{Q}[\zeta_{7 \cdot 13}]$ be the field of fractions of R.

(c) The group algebras $\mathbb{F}G$ and $\mathbb{F}H$ are isomorphic.

Because of (a) the algebras $\mathbb{F} \otimes \Gamma_G$ and $\mathbb{F} \otimes \Gamma_H$ are isomorphic and

$$\mathbb{F} \otimes \Lambda_G = (\mathbb{F} \otimes K)_3 \simeq \prod_1^{24}(\mathbb{F})_3 \simeq (\mathbb{F} \otimes L)_3 = \mathbb{F} \otimes \Lambda_H.$$

(d) Because of (c) the group rings $R_\wp G$ and $R_\wp H$ are isomorphic for every prime ideal $\wp$ of R, which does not divide 3, 7 or 13.

(e) Let $\wp$ be a prime ideal of R lying above 3. Then the exact sequence

$$0 \to I_{R_\wp}(A \times B) \cdot G \to R_\wp G \to R_\wp C \to 0$$

is two sided split, and $I_{R_\wp}(A \times B) \cdot G =: \Lambda_{G,p}$ consists of blocks of defect zero (VIII.6). Because of (d) we conclude $\Lambda_{G,p} \simeq \Lambda_{H,p}$, and so the group rings are isomorphic.

(f) Let $\wp$ be a prime ideal of R lying above 13.

Then the exact sequence

$$0 \to I_{R_\wp}(A) \cdot G \to R_\wp G \to R_\wp G/A \to 0$$

is two sided split. By the above, $G/A \simeq G/B$, and so it remains to show, that $I_{R_\wp}(A) \cdot G \simeq I_{R_\wp}(B) \cdot G$. But this is again a consequence of Clifford theory: Let M be a faithful irreducible RA-module, then its inertia group does not contain $< c >$ (cf. VIII.16) and so by Chapter XIII $I_{R_\wp}(A) \cdot G \simeq I_{R_\wp}(B) \cdot G$. Thus the group rings are isomorphic.

(g) The case of a prime ideal $\wp$ of R lying above 7 is done similarly. This completes the proof of Lemma VIII.14.

□

Lemma VIII.11 also shows

VIII.17. Lemma. *The cohomology rings of the groups G and H from Lemma VIII.11 have isomorphic cohomology rings, though the p−adic group rings are not isomorphic.*

PROOF: Since for a solvable group G the principal p−block $B_{o,G,p}$ is isomorphic to $\hat{\mathbb{Z}}_p G/O_{p'}(G)$, where $O_{p'}(G)$ is the largest normal subgroup of G with order prime to p, we only need to verify, that for the groups in Lemma VIII.11, $G/O_{p'}(G)$ is isomorphic to $H/O_{p'}(H)$ for $p = 3, 7, 13$; but that is obvious.   □

IX   Zassenhaus conjecture

§1   A Semi Local Counterexample

Zassenhaus conjectured in [Se; 83] that group bases in $\mathbb{Z}G$ are not only isomorphic, they are even conjugate in the group ring $\mathbb{Q}G$. Here we view $\mathbb{Z}G \subset \mathbb{Q}G$. This would have far reaching consequences. For example for the automorphism group of $\mathbb{Z}G$ this implies immediately that every normalized automorphism is central up to a group automorphism of the group base G.

The aim of this section is to construct a finite metabelian, supersolvable group G and an automorphism α of $RG = \mathbb{Z}_{\pi(G)}G$, where G is a group of order 6.720 (cf. below), such that this automorphism cannot be written as the composite of a central automorphism of RG with an automorphism induced from a group automorphism of G.

This is a counterexample to the Zassenhaus conjecture for semi local rings. However, in [R-S; 87 3] it was shown – using delicate manipulations with $K_1(RG)$ – that the above is also a global counterexample. The problem "local–global" for automorphisms – at least for central automorphisms – can be pinpointed. It should be noted that even this example in the semi local case gives a global counterexample for RG for the ring of algebraic integers R in a suitable algebraic number field K.

By the results of Fröhlich [Fr; 73], see also section VII, we have the commutative diagram with exact rows (for the definition of the various groups we refer to section VII):

$$
\begin{array}{ccccccccc}
0 & \longrightarrow & Cl(\mathbb{Z}G) & \longrightarrow & Picent(\mathbb{Z}G) & \xrightarrow{\varphi} & \prod_{p\mid |G|} Picent(\hat{\mathbb{Z}}_pG) & \longrightarrow & 0 \\
& & \uparrow & & \uparrow & & \| \wr & & \\
0 & \longrightarrow & Cl_{\mathbb{Z}G}(\mathbb{Z}G) & \longrightarrow & Outcent(\mathbb{Z}G) & \xrightarrow{\varphi'} & \prod_{p\mid |G|} Outcent(\hat{\mathbb{Z}}_pG). & &
\end{array}
$$

Though φ is surjective, it is not known, whether φ' is surjective. This is equivalent to the question, whether for an invertible bimodule M, in the genus – as invertible bimodules – of M there is always an N which is $\mathbb{Z}G$–free on the left, say (in this special example it turns out to be true).

A *group automorphism* ρ of G is said to be *central* provided it stabilizes all conjugacy classes, i.e. $\rho(g)$ and g are conjugate in G.

We shall first describe the group:

1.1. Definition. *Let* $\mathbb{F}_4 = \{0, 1, \zeta, \zeta^2\}$ *be the field with* 4 *elements. In the ring of* 3×3 *matrices over* $\mathbb{F}_4$ *we consider the following matrices:*

$$
s = \begin{pmatrix} 1 & 1 & 0 \\ 0 & 1 & 0 \\ 0 & 0 & 1 \end{pmatrix} ; t = \begin{pmatrix} 1 & 0 & 0 \\ 0 & 1 & 1 \\ 0 & 0 & 1 \end{pmatrix} ;
$$

$$u = \begin{pmatrix} 1 & \zeta & 0 \\ 0 & 1 & 0 \\ 0 & 0 & 1 \end{pmatrix}; v = \begin{pmatrix} 1 & 0 & 0 \\ 0 & 1 & \zeta \\ 0 & 0 & 1 \end{pmatrix};$$

$$c = \begin{pmatrix} 1 & 0 & 1 \\ 0 & 1 & 0 \\ 0 & 0 & 1 \end{pmatrix}; c' = \begin{pmatrix} 1 & 0 & \zeta \\ 0 & 1 & 0 \\ 0 & 0 & 1 \end{pmatrix}; c'' = \begin{pmatrix} 1 & 0 & \zeta^2 \\ 0 & 1 & 0 \\ 0 & 0 & 1 \end{pmatrix}.$$

The elements

$$s, t, u, v, c, c', c''$$

generate the group of upper triangular matrices over $\mathbb{F}_4$ with diagonal entries one, the unipotent radical, which we denote by H_0. It has order 4^3 and its centre is $Z = \{1, c, c', c''\}$.

For the group we refer to [Wall; 47, Sah; 68, J-M; 87] .

1.2. Definition. *We define now several central automorphisms of H_0:*

$$\sigma \quad : \quad H_0 \longrightarrow H_0 \quad : \quad s \longrightarrow c \cdot s$$
$$t \longrightarrow c \cdot t$$

The rest of the generators stay fixed.

$$\beta \quad : \quad H_0 \longrightarrow H_0 \quad : \quad s \longrightarrow c \cdot s$$
$$t \longrightarrow t$$

The rest of the generators stay fixed.

$$\sigma \cdot \beta \quad : \quad H_0 \longrightarrow H_0 \quad : \quad s \longrightarrow s$$
$$t \longrightarrow c \cdot t$$

The rest of the generators stay fixed.

1.3. Claim. *The automorphisms σ, β and $\sigma \cdot \beta$ are central automorphisms, which are not inner on H_0.*

PROOF OF CLAIM 1.3: We see that

$$t \cdot s \cdot t = c \cdot s, s \cdot t \cdot s = c \cdot t$$

and so a simple calculation shows that the automorphisms are central.

For proving that the automorphisms are not inner, we look at the centralizers of s, t, u and v in H_0.

$$C_{H_0}(s) = C_{H_0}(u) = \begin{pmatrix} 1 & \mathbb{F}_4 & \mathbb{F}_4 \\ 0 & 1 & 0 \\ 0 & 0 & 1 \end{pmatrix}, C_{H_0}(t) = C_{H_0}(v) = \begin{pmatrix} 1 & 0 & \mathbb{F}_4 \\ 0 & 1 & \mathbb{F}_4 \\ 0 & 0 & 1 \end{pmatrix},$$

as one sees from easy calculations. If now σ were inner, the intersection

$$s \cdot C_{H_0}(t) \cap t \cdot C_{H_0}(s) \cap C_{H_0}(u) = \emptyset$$

would be non empty since two conjugating elements of one specific generator differ by an element in its centralizer, a contradiction. For β we see that

$$C_{H_0}(u) \cap t \cdot C_{H_0}(s) = \emptyset$$

and so β cannot be inner. Since $C_{H_0}(v) \cap s \cdot C_{H_0}(t) = \emptyset$, we see that $\sigma\beta$ is not inner. q.e.d. Claim 1.3 $\square$

We now define our group, which will turn out to be a counterexample to the Zassenhaus conjecture.

1.4. Definition. *(The group)* We define three modules X, Y, Z for H_0 – written multiplicatively – on which $< u, v, c, c', c'' >$ acts trivially:

$$X =< x : x^3 = 1 > \text{ with the action } {}^s x = x^{-1} \text{ , } {}^t x = x^{-1},$$

$$Y =< y : y^7 = 1 > \text{ with the action } {}^s y = y^{-1}, {}^t y = y,$$

$$Z =< z : z^5 = 1 > \text{ with the action } {}^s z = z, {}^t z = z^{-1}.$$

We then put

$$G := (X \times Y \times Z) \rtimes H_0$$

to be the semi direct product of H_0 with the direct sum of these modules and put $H = G/(Y \times Z)$. Then G has order 6.720.

The automorphisms σ, β and $\sigma \cdot \beta$ extend to automorphisms of G, by letting them act as identity on $X \times Y \times Z$ as is easily seen. By abuse of language we shall denote the extended automorphisms also by σ, β and $\sigma \cdot \beta$ respectively.[27]

1.5. Claim. σ *is a central automorphism on* $X \rtimes H_0$, *but there are no central automorphisms* ρ_Y^σ *and* ρ_Z^σ *of* $(X \times Y) \rtimes H_0$ *and* $(X \times Z) \rtimes H_0$ *resp. with*

$$\rho_Y^\sigma = \rho_Z^\sigma \cdot \sigma$$

induced on $X \rtimes H_0$.

β *is a central automorphism on* $Y \rtimes H_0$, *but there are no central automorphisms* ρ_X^β *and* ρ_Z^β *of* $(X \times Y) \rtimes H_0$ *and* $(Y \times Z) \rtimes H_0$ *with*

$$\rho_X^\beta = \rho_Z^\beta \cdot \beta$$

induced on $Y \rtimes H_0$.

$\sigma \cdot \beta$ *is a central automorphism on* $Z \rtimes H_0$, *but there are no central automorphisms* $\rho_X^{\sigma \cdot \beta}$ *and* $\rho_Y^{\sigma \cdot \beta}$ *of* $(X \times Z) \rtimes H_0$ *and* $(Y \times Z) \rtimes H_0$ *with*

$$\rho_X^{\sigma \cdot \beta} = \rho_Y^{\sigma \cdot \beta} \cdot \sigma \cdot \beta$$

induced on $Z \rtimes H_0$.

---

27   We point out, that this group stands for a whole family of groups. One can even take X and Y of order 3 and Z of order 5

PROOF OF CLAIM 1.5: We only prove the first part since the others are done similarly.

We hence assume the existence of the central automorphisms ρ_Z^σ and ρ_Y^σ. We do have $H^1(H_0, Y) = H^1(H_0, Z) = 0$ since the orders of the groups are relatively prime. Because of the Schur Zassenhaus Theorem[28] we may vary both automorphisms by an inner automorphism so that ρ_Z^σ and ρ_Y^σ stabilize H_0. If we conjugate by s or t or st we get that ρ_Y^σ is the identity on $X \times Y$ and by modifying with another inner automorphism we can arrange that u and v are stabilized by ρ_Y^σ. The latter one is obtained as follows: We have $\rho_Y^\sigma(u) = uz, \rho_Y^\sigma(v) = vz'$ for central elements z, z' of H_0. Since u and v centralize x and y, $\rho_Y^\sigma(u)$ and u must be conjugate in $C_{H_0}(x) \cap C_{H_0}(y) =< u, v, Z(H_0) >$. So we see that $\rho_Y^\sigma(u) \in \{u, uc''\}$. The same conclusion yields $\rho_Y^\sigma(v) \in \{v, vc''\}$. If $\rho_Y^\sigma(u) = uc''$, we can conjugate by v fixing x, y and v. Similar observations yield to the hypothesis that ρ_Y^σ stabilize u and v. So we must have $\rho_Y^\sigma(s) \in \{s, sc'\}$ and $\rho_Y^\sigma(t) \in \{t, tc'\}$. That is since $\rho_Y^\sigma(s)$ must be conjugate to s in $C_{H_0}(y)$ and $\rho_Y^\sigma(t)$ must be conjugate to t in $C_{H_0}(x)$.

We now use the hypotheses $\rho_Y^\sigma = \rho_Z^\sigma \sigma$ to determine ρ_Z^σ.

$$\rho_Z^\sigma(s) \in \{sc, sc''\}, \rho_Z^\sigma(t) \in \{tc, tc''\}, \rho_Z^\sigma(u) = u, \rho_Z^\sigma(v) = v \text{ and } \rho_Z^\sigma(z) \in \{z, z^{-1}\}.$$

If ρ_Z^σ does not stabilize z we can conjugate vz to its image only by elements in $sC_{H_0}(z)$, but they conjugate v to vc or to vc'. Otherwise $C_{H_0}(z)$ conjugates t to tc' or to t and hence tz is not conjugate to its image under ρ_Z^σ. This contradiction proves the claim. q.e.d.   □

Let R_0 be the ring of algebraic integers of the algebraic number field K, which we assume to be a splitting field for G and all of its subgroups. Let R be a semi localization of R_0, such that $2, 3, 5, 7$ are not invertible in R.

1.6. Remark. We shall construct an augmented automorphism α of RG, which will be a counterexample to the Zassenhaus conjecture. In order to do so, we shall construct local automorphisms $\alpha_\wp$ for every $\wp \in max(R)$ of $R_\wp G$. The local data $\alpha_\wp$ give then rise to an automorphism α of RG if and only if the automorphisms $\alpha_{\wp_1} \cdot \alpha_{\wp_2}^{-1}$ of KG are central for all $\wp_1, \wp_2 \in spec(R)$. This follows easily from the interpretation of automorphisms as invertible bimodules (cf. Section VII § 1).

In fact: If $\alpha_{\wp_1} \alpha_{\wp_2}^{-1}$ is a central automorphism, then

$$M_i = {}_{\alpha_{\wp_i}} \mathbb{Z}_{\wp_i} G_1 \subseteq {}_{\alpha_{\wp_1}} \mathbb{Q} G_1,$$

and we can form

$$M := \bigcap_i {}_{\alpha_{\wp_i}} \mathbb{Z}_{\wp_i} G_1,$$

---

28   The Schur Zassenhaus theorem states that $H^2(G, A) = 0$ if A is a finite abelian group with $(|G|, |A|) = 1$.

which is free from the right, and thus gives rise to an automorphism. The converse is obvious.

The construction presented here is totally different from the original construction, where we were very carefully analyzing the group ring as being built from various pieces, which were represented as pullbacks. On each of these we did construct semi local automorphisms, which we then carefully modified, so that they would fit together.

1.7. Theorem.

(a) *If* $5 \notin \wp(\neg 5)$ *for some* $\wp(\neg 5) \in max(R)$, *then the group ring* $R_{\wp(\neg 5)}G$ *decomposes as follows:*

$$R_{\wp(\neg 5)}G = I_{\wp(\neg 5)}(Z) \cdot G \ \times \ R_{\wp(\neg 5)}[(X \times Y) \cdot H],$$

where $I_{\wp(\neg 5)}(Z) \cdot G$ *is the augmentation ideal of* Z *in* $R_{\wp(5)}G$. *We then define* $\alpha_{\wp(\neg 5)}$ *on the two pieces:*

$$\alpha_{\wp(\neg 5)} \ : \ R_{\wp(\neg 5)}[(X \times Y) \cdot H] \longrightarrow R_{\wp(\neg 5)}[(X \times Y) \cdot H]$$

is σ.

$$\alpha_{\wp(\neg 5)} \ : \ I_{\wp(\neg 5)}(Z) \cdot G \longrightarrow I_{\wp(\neg 5)}(Z) \cdot G$$

is the identity.

(b) *If* $5 \in \wp(5)$ *for some* $\wp(5) \in max(R)$, *then the group ring* $R_{\wp(5)}G$ *decomposes as follows:*

$$R_{\wp(5)}G = I_{\wp(5)}(Y) \cdot G \ \times \ R_{\wp(5)}[(X \times Z) \cdot H].$$

We then define $\alpha_{\wp(5)}$ *on the two pieces:*

$$\alpha_{\wp(5)} \ : \ I_{\wp(5)}(Y) \cdot G \longrightarrow I_{\wp(5)}(Y) \cdot G$$

is σ.

$$\alpha_{\wp(5)} \ : \ R_{\wp(5)}[(X \times Z) \cdot H] \longrightarrow R_{\wp(5)}[(X \times Z) \cdot H]$$

is the identity.

Then the local data $\alpha_{\wp}$, $\wp \in spec(R)$ *give rise to an automorphism of* α *of* RG, *which is a counterexample to the Zassenhaus conjecture.*

PROOF OF THEOREM 1.7: *We first show, that the local data give rise to an automorphism of* RG:
For this we have to look at the intersection of the various pieces:

1.8. Claim. σ *is central on*

$$K[(X \times Y) \cdot H] \ \cap \ K[(X \times Z) \cdot H]$$

PROOF OF CLAIM 1.8: The above intersection is just $K[X \cdot H]$, on which σ is central (cf. 1.5).

q.e.d.   □

1.9. Claim. σ *is* central *on* $KI(Y) \cdot G \cap KI(Z) \cdot G$

PROOF OF CLAIM 1.9: Let V be an irreducible module in the above intersection, then V is an irreducible KG–module, on which neither Y nor Z act trivially.

We now invoke Clifford's theory (Section XIII): Let $\chi_{Y,Z}$ be an irreducible character which is neither trivial on Z nor on Y – note that Y and Z are abelian and that K is a splitting field for Y and Z. Then the inertia group $I_{\chi_{Y,Z}}$ in KG is

$$I_{\chi_{Y,Z}} = (X \times Y \times Z) \rtimes < u, v, c, c', c'' > .$$

σ centralizes $\chi_{Y,Z}$ and acts trivially on $I_{\chi_{Y,Z}}$. With Theorem XIII.2 we conclude, that $^\sigma V \simeq V$. This happens since Clifford's theorem provides a Morita equivalence between the modules over the inertia group and those over G. And thus σ is central on $KI(Y) \cdot G \cap KI(Z) \cdot G$. q.e.d.   □

Because of Remark 1.6 the local automorphisms constructed above give rise to an automorphism α of RG.

1.10. Claim. *The automorphism α is a counterexample to the Zassenhaus conjecture; i.e. it cannot be written as a central automorphism γ followed by an automorphism induced by a group automorphism ρ.*

PROOF OF CLAIM 1.10: Because of the class sum correspondence (cf. Theorem IV.1), α sends class sums to class sums, and we have to show, that this cannot be achieved by a group automorphism.

Since Z centralizes $s \cdot x \cdot y$, all G–conjugates of $s \cdot x \cdot y$ are the G/Z–conjugates of $s \cdot x \cdot y$ in $(X \times Y) \cdot H$. Furthermore the *conjugacy class* of $s \cdot x \cdot y$ is mapped under α to the conjugacy class of $s \cdot c \cdot x \cdot y$ – note that $s \cdot x \cdot y$ and $s \cdot c \cdot x \cdot y$ are not G–conjugate.

On the other hand, if χ_Z is a non trivial irreducible character of Z, then the inertia group of χ_Z is

$$I_{\chi_Z} = (X \times Y \times Z) \rtimes < s, u, v, c, c', c'' > .$$

Assume now that

$$\alpha = \gamma \cdot \rho \, , \gamma \text{ a central automorphism}, \rho \text{ a group automorphism}.$$

Since α is central on the components of KG which come from χ_Z – in the sense of Clifford's theory – we must have

$$^\rho \chi_Z = \,^{g_0} \chi_Z \text{ for a } g_0 \in G,$$

since, if we restrict the G module V corresponding to the character χ_Z to I_{χ_Z} by Clifford's theory, we get a direct sum of conjugate modules.

But then $\rho \cdot$ (conjugation by g_0) must act as an inner automorphism on I_{χ_Z}/Z, a contradiction, since α sends $s \cdot x \cdot y$ to a conjugate of $s \cdot c \cdot x \cdot y$. q.e.d.   □

But now the Theorem is proved.   □

1.11. Remark. Similar constructions can be made with β and $\sigma \cdot \beta$.

1.12. Remark. It should be noted that the above automorphisms $\alpha_\wp$ are defined entirely in terms of group automorphisms, and hence we can define them also over $\hat{\mathbb{Z}}_p G$, noting that the decompositions of the group ring (cf. 1.7) $R_\wp G$ occur already over $\hat{\mathbb{Z}}_p G$.
Thus we have a counterexample $\alpha_{\mathbb{Z}}$ to the Zassenhaus conjecture for $\mathbb{Z}_{\pi(G)} G$.

Let us reflect for a moment on a *global construction of a counterexample to the Zassenhaus conjecture*, for $\mathbb{Z}G$ as it is worked out in [R-S; 87 3]:

Using the techniques of invertible bimodules (Section VII) – Fröhlich's localization sequence (cf. [Fr; 73],[C-R2; 87]), there exists an invertible bimodule M for $\mathbb{Z}G$ for the very same G, which localizes to $\alpha_{\mathbb{Z}}$. The problem now is, to find such an M, which is $\hat{\mathbb{Z}}_p G$–free on the left as $\hat{\mathbb{Z}}_p G$–module. Whether this is always possible seems to be an open problem (cf. below (section VII §1) and problem 1.13); however, in the above special situation, we were able to construct such a module ([R-S; 87 3]), using some kind of Mayer–Vietoris sequence for invertible bimodules; of importance here is, that $\mathbb{Z}G$ satisfies the Eichler condition [Re; 75]. If one is less ambitious, one can argue as follows:

Choose a suitably large algebraic number field with ring of integers S such that $S \otimes_{\mathbb{Z}} M$ is free as left SG–module. Then $S \otimes_{\mathbb{Z}} M$ gives rise to an automorphisms α_S of SG, which is a counterexample to the Zassenhaus conjecture for SG.

1.13. Problem.   Let $M(p)$, $p \in spec(\mathbb{Z})$ be a family of invertible $\hat{\mathbb{Z}}_p G$–bimodules, which are isomorphic as $\mathbb{Q}G$-modules. Is it possible, to find an invertible bimodule M for $\mathbb{Q}G$, which is free as left $\mathbb{Z}G$-module, such that $\hat{\mathbb{Z}}_p \otimes_{\mathbb{Z}} M \simeq M(p)$ for all $p \cdot \mathbb{Z} \in spec(\mathbb{Z})$?
Equivalently, does every genus of invertible bimodules contains one, which is free as left $\mathbb{Z}G$-module?

Klingler used the above ideas ([R-S; 87 3]) and pullback–constructions to give the automorphism constructed above explicitly ([Kl; 91]).

§2   A general observation on the Zassenhaus conjecture

Let us add some general remarks about the Zassenhaus conjecture for solvable groups. Let G be solvable. Because of Theorem VI.1 the Zassenhaus conjecture holds for all the groups $G/O_{p'_i}(G)$ for all primes p_i which divide $|G|$. Moreover, there is an injective homomorphism

$$\phi \; : \; G \longrightarrow \prod_{i=1}^{n} (G/O_{p_i'}(G);$$

i.e., G is a subdirect product of groups, for which the Zassenhaus conjecture holds. We point out, that the groups $O_{p_i'}(G)$ are characteristic in G, and moreover, the induced augmentation ideal $I(O_{p_i'}(G)) \cdot \mathbb{Z}G$ is determined by $\mathbb{Z}G$ by the normal subgroup correspondence (V.1).

As a general setup we now assume:

2.1. Assumption. Let G be a finite group with normal subgroups $\{K_\alpha\}_{\alpha \in I}$, where I is a finite index set, such that

- $\cap_{\alpha \in I} K_\alpha = 1$
- the augmentation ideals $I(K_\alpha) \cdot \mathbb{Z}G$ are determined by $\mathbb{Z}G$.

We then let

$$\phi_\alpha : G \longrightarrow G/K_\alpha =: G_\alpha$$

be the natural projection, and we put for a subset $S \subset I$

$$\phi_S : G \longrightarrow G/(\prod_{\alpha \in S} K_\alpha) =: G_S.$$

Then G is the projective limit of G_S over the subsets S of I, i.e. $G = lim.proj.(G_S)$ (cf. Section II). However, conjugacy classes and group rings do not behave well under projective limits, and so we have for the conjugacy classes in general

$$Cl(G) \neq lim.proj(Cl(G_S)) =: X = \{X_g\}_{g \in G},$$

and for the group rings in general

$$\mathbb{Z}G \neq lim.proj.(\mathbb{Z}G_S) =: \Gamma.$$

If we *assume* that the Zassenhaus conjecture holds for the groups G_S, then a weaker form of the Zassenhaus conjecture, which is adapted to projective limits, is the following:

2.2. Conjecture. Weak form of the Zassenhaus conjecture (with respect to the family of subgroups G_α).

If $\mathbb{Z}G = \mathbb{Z}H$, then there exists an automorphism ρ from G to H, which is an automorphism over X. (Note that because of IV.1 $X_g = X_h$ for a suitable bijection between G and H.) We could even replace $\mathbb{Z}G$ by Γ.

Let us look at the special case, where $I = \{1, 2\}$ – this is also the case for the counterexample to the Zassenhaus conjecture explained above. Then G is the pullback

$$
\begin{array}{ccc}
G & \longrightarrow & G_1 \\
\downarrow & & \downarrow \\
G_2 & \longrightarrow & G' \; .
\end{array}
$$

We assume now, that $\sigma : \mathbb{Z}G \longrightarrow \mathbb{Z}G$ is an augmented automorphism. Because of 2.1 σ induces augmented automorphisms $\sigma_i : \mathbb{Z}G_i \longrightarrow \mathbb{Z}G_i, i = 1, 2$ and $\sigma' : \mathbb{Z}G' \longrightarrow \mathbb{Z}G'$. According to the hypothesis, the Zassenhaus conjecture is true for G_1, G_2 and G'; i.e.,

$$\sigma_i|_{\mathbb{Z}G_i} = conj(a_i) \cdot \rho_i, \ i = 1, 2 \ ,$$

where a_i is a unit in $\mathbb{Q}G_i$ stabilizing $\mathbb{Z}G_i$, and $\rho_i \in Aut(G_i)$, $i = 1, 2$. Note however, that neither ρ_i nor $conj(a_i)$ is uniquely determined. They are only unique up to central automorphisms: a central automorphisms γ_i of G_i – i.e. $\gamma_i(x)$ and x are conjugate in G_i for $x \in G_i$ – can also be written as $conj(c_i)$, $c_i \in \mathbb{Q}G_i$.

On $\mathbb{Z}G'$ we have $\sigma_1 = \sigma_2$; thus $conj(a_1) \cdot \rho_1 = conj(a_2) \cdot \rho_2$ on $\mathbb{Z}G'$, and so

$$\rho_1 \cdot \rho_2^{-1} = conj(a_1^{-1}) \cdot conj(a_2) =: \gamma_{12}$$

on $\mathbb{Z}G'$. Thus γ_{12} is a *central group automorphism* of $\mathbb{Z}G'$; note that though the above construction is not unique, the automorphism γ_{12} is independent of the chosen "Zassenhaus decomposition".

2.3. Proposition. *If $\gamma_{12} = \gamma_1 \cdot \gamma_2^{-1}$ for central group automorphisms γ_i of $G_i, i = 1, 2$, then the Zassenhaus conjecture is true for the pullback Γ:*

$$
\begin{array}{ccc}
\Gamma & \longrightarrow & \mathbb{Z}G_1 \\
\downarrow & & \downarrow \\
\mathbb{Z}G_2 & \longrightarrow & \mathbb{Z}G'
\end{array}
\quad ;
$$

i.e. $\sigma = conj(a) \cdot \rho$ on Γ for a group automorphism ρ of G and a unit a in $\mathbb{Q}\Gamma$, centralizing Γ, and conversely.

2.4. Remark.

- The proof of the above lemma consists of straight forward computations.
- From the above Lemma it follows, that the Zassenhaus conjecture holds for Γ, if every central automorphism of G' is an inner automorphism, or if it can be written as a product of central automorphisms of G_1 and G_2.
- We note that the above condition $\gamma_{12} = \gamma_1 \cdot \gamma_2^{-1}$ is the condition that a 1–cocycle is a 1–coboundary, in a Čech style cohomology theory, which is analogous to the Čech 1–cohomology of a topological space with respect to a cover by two open sets.
- For projective limits as above, the validity of the weak form of the Zassenhaus conjecture for Γ is equivalent to the condition, that a Čech style 1–cocyle is a 1–coboundary.
- The counterexample of the Zassenhaus conjecture in paragraph 1 is also a counterexample to the weak form of the Zassenhaus conjecture.

X   Variations of the Zassenhaus conjecture

by W.Kimmerle

Throughout this section R is an integral domain of characteristic zero, G a finite group and no prime divisor of $|G|$ is invertible in R . K denotes a field containing R. Therefore by the previous sections we have between group bases in RG a class sum correspondence.

Since the Zassenhaus-conjecture (cf. section IX) is in general not valid, it is necessary to discuss replacements, which possibly still imply a positive solution of the isomorphism problem for integral group rings. The first object of this section is to establish some equivalent formulations of the Zassenhaus conjecture. The first one is its well known formulation in terms of the class sum correspondence.

X.1. Lemma. *Let R be an integral domain of characteristic zero, let G be a finite group and let H be a group basis of RG. Then the following are equivalent.*
1) H is conjugate by a unit of KG to G within RG, where K denotes some field containing R.
2) There exists a class sum correspondence $\sigma : G \to H$, which is a group isomorphism between G and H.

PROOF: Class sums are central elements. So conjugation fixes such elements. Hence obviously 1) implies 2).

Conversely the Noether-Skolem Theorem [C-R1; 82, §3E] implies that an isomorphism between group bases of RG fixing the centre of RG, extended to KG, is given by an inner automorphism of KG.   □

§1   Automorphisms

In the next step we prove a link between the Zassenhaus conjecture and properties of normalized automorphisms of RG. Note that the Zassenhaus conjecture a priori makes the statement that all group bases are conjugate in the total ring of quotients, i.e. in particular that all group bases are isomorphic and thus the isomorphism problem for RG has a positive solution.

A ring **automorphism** σ of RG is called **normalized**, if σ preserves augmentation. The normalized ring automorphisms of RG form a normal subgroup $Aut_n(RG)$ of the group of all ring automorphisms of RG. The latter one we denote by $Aut(RG)$. Clearly $Aut(RG)$ is a direct product of $Aut_n(RG)$ and R^*. Thus no information is lost, if one studies only $Aut_n(RG)$.

If B is a group basis of RG, then each group automorphism α may be R-linearly extended to a normalized ring automorphism of RG. We call such ring automorphisms induced from α.

1.1. Lemma. *Consider the following two statements.*

(1) Let $\sigma \in Aut_n(RG)$ or $\in Aut_n(R(G \times G))$. Let H be a group basis of RG or of $R(G \times G)$. Then σ may be written as the composition of a ring automorphism induced from a group automorphism of H and a ring automorphism τ which has the property that τ fixes for each normal subgroup N of G, $G \times G$ rsp. the sum of the elements of N. (We call this sum as usual the trace of N).

(2) If X and Y are group bases of RG, then there exists an isomorphism σ between X and Y such that σ fixes for each normal subgroup N of G the sum of the elements of N.

Then (1) implies (2).

Note that Lemma 1.1 reduces the question of the isomorphism problem to a question about normalized ring automorphisms.

PROOF: Let H be a group basis of RG. It suffices to prove (2) for $X = G$ and $Y = H$. Identify RG with the subrings $R(G \times 1), R(1 \times G)$ rsp. of $R(G \times G)$.

Hence we obtain in these subrings group bases H_1, H_2 rsp. which are isomorphic to H. Then obviously $B_1 = < H_1, 1 \times G >, B_2 = < H_2, G \times 1 >$ rsp. are isomorphic group bases of $R(G \times G)$.

By (1) there exists $\sigma \in Aut_n(R(G \times G))$ with $\sigma(B_1) = B_2$ and σ fixes the traces of the normal subgroups of $G \times G$. By Theorem V.1 the automorphism σ fixes also the traces of the normal subgroups of B_1, in particular that one of H_1. However looking at the trivial submodule of $R(G \times 1)$ we see that the trace of H_1 coincides with that one of $G \times 1$. Hence σ must map H_1 to $G \times 1$ and therefore G and H are isomorphic. The restriction of σ to $R(G \times 1)$ must then be an automorphism.

By (1) this restriction may be modified by a group automorphism to an automorphism τ such that τ fixes the traces of all normal subgroups of $G \times 1$. $\quad\square$

1.2. Remarks.

a) Lemma 1.1 shows that the Zassenhaus conjecture holds for RG provided two isomorphic group bases in RG and in $R(G \times G)$ are conjugate by a unit in KG. For, if isomorphic group bases are conjugate, there is a normalized automorphism between them fixing the class sums and therefore for each normal subgroup the sum of its elements. By Lemma 1.1 it follows that group bases of RG are always isomorphic. But by assumption such group bases are conjugate and so the Zassenhaus conjecture holds for RG.

So, if one can prove for a class of finite groups which is closed under direct products that each normalized ring automorphism is with respect to each group basis B the composition of a ring automorphism induced from a group automorphism of B and an inner automorphism of KG, then the conjecture of Zassenhaus is established only by properties of $Aut(RG)$.

b) In order to apply Lemma 1.1 it is however not necessary that normalized group ring automorphisms modulo group automorphisms fix **all** class sums. In particular we consider in the sequel the following two variations.

1.3. ZC-Variation 1. *Let X and Y be group bases of RG. Then there exists an isomorphism $\sigma : X \to Y$, such that σ fixes the class sums of all cyclic subgroups of G (for the definition of such class sums see the following remarks).*

1.4. ZC-Variation 2. *Let X and Y be group bases of RG. Then there exists an isomorphism $\sigma : X \to Y$ such that σ fixes the class sums of all elements of prime power order.*

1.5. Remarks.

a) Ad **Variation 1.** By a **class sum of a conjugacy class** of a cyclic subgroup we mean the following. Let Z be a cyclic subgroup, then for each $g \in Z$ which is not contained in a maximal subgroup of G (or in other words for each generating element of Z) denote by $C(g)$ its conjugacy class in G. Let $U(Z)$ be the union of these conjugacy classes. Then the group ring element K_Z, which is the sum of the elements of $U(Z)$, is called the class sum of the conjugacy class of Z (clearly K_Z and K_{Z^g} coincide for each $g \in G$).

Now note that, if a normalized ring automorphism of RG fixes each class sum of cyclic subgroups, then it fixes for each normal subgroup N the sum of its elements. Consequently we can apply Lemma 1.1.

b) Ad **Variation 2.** Each finite group X may be written as a product

$$P_1 \cdot ... \cdot P_n,$$

where P_i denotes the union of the Sylow p_i-subgroups and $\pi(X) = \{p_1, ..., p_n\}$. Let N be a normal subgroup of G and assume that $\sigma \in Aut_n(RG)$ fixes the class sums of the conjugacy classes of all elements of prime power order, then it follows that the product over all sums of elements of prime power order which are contained in N is of the form

$$x = \Sigma_i \alpha_i \cdot K_i \text{ with } \alpha_i \in I\!N \text{ and } N = \cup C_i,$$

where K_i denotes the class sums in N and C_i denotes the corresponding conjugacy classes. Clearly σ fixes x. Let $y = \Sigma_{n \in N} n$. Clearly $y \cdot x = \epsilon(x) \cdot y$, where ϵ denotes the augmentation map. By Theorem V.1 $\sigma(y) = \Sigma_{m \in M} m$, where M is a normal subgroup of G with $|M| = |N|$. Now

$$\sigma(y \cdot x) = \sigma(y) \cdot x = \epsilon(x) \cdot \sigma(y)$$

and the last equality is only possible, if each summand of x is an element of M (note R has characteristic zero). This shows that $M = N$ and that $\sigma(y) = y$. So we are again in the position to apply Lemma 1.1.

1.6. Aut-Variations. *Suppose that each normalized automorphism τ of RG may be written with respect to each group basis H as*

$$\tau = \alpha \cdot \sigma,$$

where α is induced from a group automorphism of H and σ is a normalized automorphism of RG such that σ fixes all class sums of conjugacy classes of cyclic subgroups of G. Then we say that Aut-Variation 1 holds. If σ fixes all class sums of conjugacy classes of elements of prime power order, then we say that Aut-Variation 2 holds.

By Remarks 1.5 and by Lemma 1.1 we obtain the following Corollaries.

1.7. Corollary. *Let C be a class of finite groups closed under products. Then for $i = 1,2$ the following are equivalent.*

1) ZC – Variation i holds for all $G \in C$.

2) Aut – Variation i holds for all $G \in C$.

1.8. Remarks.

a) Assume that $R = \mathbb{Z}$ and $K = \mathbb{Q}$. Then ZC–Variation 1 holds if, and only if, each normalized group automorphism τ fixes each Wedderburn component of $\mathbb{Q}G$.

There is no group known which does not have this property. All counterexamples to the Zassenhaus conjecture constructed by Roggenkamp and Scott (cf. section IX) fix the components of $\mathbb{Q}G$.

b) Variations 1 and 2 might hold independently of each other.

§2   Indecomposable groups

For some special groups G it suffices to consider only the action of $Out(G)$ on the conjugacy classes in order to establish the Zassenhaus conjecture. The basic observation is the following. Let $\sigma \in Aut_n(RG)$. Then σ permutes the class sums, i.e. σ induces a permutation, also denoted by σ, of the conjugacy classes of G. By Theorem V.1 the automorphism σ must preserve the sizes of the conjugacy classes, the order of representatives of C and $\sigma(C)$ and the power map on the classes. Furthermore it must be compatible with the normal subgroup structure, *i.e.* if the class C is contained in $N = \cup C_i$, then $\sigma(C)$ must be contained in the normal subgroup $N^* = \cup \sigma(C_i)$.

For a finite group G denote by $Cl(G)$ the set of its conjugacy classes. Let C_1 , $C_2 \in Cl(G)$. Define $C_1 \le C_2$ if , and only if, there exists a natural number n such that $x^n \in C_2$ for some $x \in C_1$. Obviously this gives $Cl(G)$ the structure of a poset.

Let $Aut(Cl(G))$ be the group of poset automorphisms of $Cl(G)$ preserving the order of the representatives of associated classes, their size and the normal subgroup structure. Now by the remarks above $Aut_n(RG)$ is a subgroup of $Aut(Cl(G))$. Clearly the central normalized ring automorphisms act trivially on $Cl(G)$. On the other hand $Out(G)$ yields naturally a subgroup of $Aut(Cl(G))$. We denote this subgroup also by $Out(G)$. Hence we obtain immediately the following proposition.

2.1. Proposition. *If $Aut(Cl(G)) = Out(G)$, then each $\sigma \in Aut_n(RG)$ may be written as the composition of a group automorphism of G followed by a central automorphism.*

2.2. Examples.

a) Let $G = A_6$. Note that in contrast to all other alternating groups $Out(A_6)$ does not have order 2, in fact $Out(A_6)$ is a Kleinian four group, *cf.* [Atl; 85] or [Hu; 67, ch.II,5.5].

We use the cycle type of the permutations to describe the conjugacy classes. $Cl(A_6)$ has the following form.

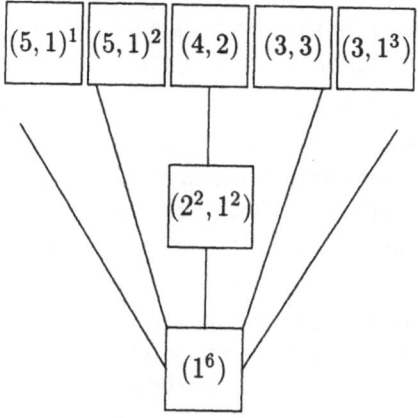

The length of the classes $(5, 1)^1$ and $(5, 1)^2$ coincides as well as those of $(3, 1^3)$ and $(3, 3)$. All other classes will be fixed by the action of $Aut(Cl(A_6))$. Consequently $|Aut(Cl(A_6))| \leq 4$.

On the other hand conjugation with a transposition of S_6 interchanges (5,1) and (5,2) and fixes all the other classes. Furthermore each outer automorphism of S_6 interchanges the classes (3,3) and (3,1,1,1). Thus $Out(A_6)$ acts faithfully on the poset and consequently each normalized automorphism of RA_6 may be written as a product of a group automorphism followed by a conjugation with a unit of KA_6. Moreover the Zassenhaus conjecture holds for RA_6, since the isomorphism problem for RA_6 has a positive solution. The last statement follows from the fact that $RA_6 \cong RH$ implies that H is simple and that A_6 is the only simple group of order 6!/2.

b) $Cl(S_6)$ has the form

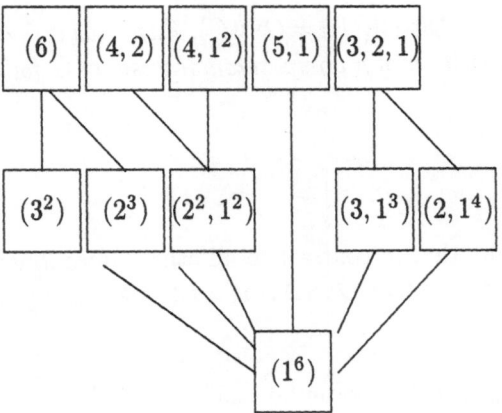

An outer automorphism of $\mathbb{Z}S_6$ interchanges the classes (6) and (3,2,1) and the subposets sitting beyond these classes. There is no $\sigma \in Aut(Cl(S_6))$ with $\sigma((4,2)) = (4,1^3)$, as (4,2) belongs to A_6 whileas $(4,1^2)$ does not. Thus we get similarly to example a) that the Zassenhaus conjecture holds for S_6, as the isomorphism problem is valid (e.g. by Theorem 2.2).

Only for some non abelian simple groups the Zassenhaus conjecture has been established so far. However with respect to symmetric groups the situation is different. In [Pe; 76] it is shown that normalized automorphisms of $\mathbb{Z}S_n$ may be written as group automorphisms followed by central automorphisms. The proof depends only on the class sum correspondence. So it applies also in the more general situation with R instead of $\mathbb{Z}$.

We remark that Peterson called a group G for which $\mathbb{Z}G$ has this property an E.R. group, E.R. stands for elementary represented. The series of the symmetric groups was one of the first ones in the area around the Zassenhaus conjecture. Note that the Zassenhaus conjecture and Peterson's result both date to the mid–seventies.

That the isomorphism problem for symmetric groups has a positive solution may be proved by different methods. Theorem 2.2 is one possibility, character tables another one, cf. [San; 85]. Our next aim is to show that our approach via the reduction to automorphisms implies this automatically.

2.3. Proposition. *Assume that each group basis Y of $R(G \times H)$ has a decomposition $Y = Y_1 \times Y_2$ such that the following holds. Each group basis X of RY_1, which is isomorphic to a factor of a product decomposition of Y, is isomorphic to Y_1.*

Then for $i = 1, 2$ Aut–Variation i is valid for $R(G \times H)$ provided Aut– Variation i holds for RY_1 and RY_2. Moreover isomorphic group bases of $R(G \times H)$ may be mapped into each other by a central automorphism provided this is the case for RY_1 and RY_2.

PROOF: Let $\sigma \in Aut_n(R(G \times U))$ and let Y be a group basis of $R(G \times U)$. Note that the normal subgroup correspondence shows that Y is a direct product.

By assumption we have a decomposition $Y = Y_1 \times Y_2$ with special properties. Denote by X_i the normal subgroup correspondents of $\sigma(Y_i)$ in Y. Because we have for the augmentation ideals $I(\sigma(Y_i)) \cdot R(G \times U) = I(X_i) \cdot R(G \times U)$, we get that X_i is isomorphic to a group basis of RY_i. On the other hand we know that $Y \cong X_1 \times X_2$. By assumption we conclude that $X_1 \cong Y_1$ and cancellation gives $X_2 \cong Y_2$.

Consequently we can modify σ by a group automorphism of Y to a ring automorphism τ such that τ fixes $I(Y_i) \cdot R(G \times U)$. Hence τ induces automorphisms τ_i of $R(G \times U/Y_i)$. By assumption , accordingly to which Aut-variation is valid for Y_i, τ_i may be modified by group automorphisms α_i such that $\alpha_i \cdot \tau_i$ fixes the desired class sums. Clearly $\alpha_1 \times \alpha_2$ is then an automorphism of Y which permits the modification of τ to a ringautomorphism ρ fixing the class sums of the considered Aut-variation belonging to Y_i. However Y_1 and Y_2 centralize each other and an arbitrary class sum of a conjugacy class of cyclic subgroups is a product of class sums of this type of Y_1 and Y_2. The same is true for class sums of elements of prime power order. Therefore ρ is a ring automorphism with $\rho(Y) = \sigma(Y)$ and ρ fixes all class sums of the desired type.   □

2.4. Corollary. *The class of finite groups satisfying the Zassenhaus conjecture is closed under direct products.*

The same holds with respect to ZC-variations 1 and 2.

PROOF: It is an immediate consequence of the normal subgroup correspondence that the isomorphism problem has for $G \times H$ a positive solution provided this is the case for G and H. Hence it suffices to establish the suitable Aut-variation. Now let Y be a group basis of $R(G \times H)$. Again by the normal subgroup correspondence we get that $Y = Y_1 \times Y_2$ with $RY_1 \cong RG$ and $RY_2 \cong RH$. Hence obviously the assumptions of Proposition 2.3 is satisfied, if the Zassenhaus conjecture or one of its variations holds for RG and RH. Thus Proposition 2.3 yields the result.   □

We call a finite group G **indecomposable**, if $G \cong A \times B$ always implies $A = 1$ or $B = 1$. Since in this sense the Krull-Schmidt Theorem holds for finite groups, it follows that for an indecomposable group G a group basis Y of $R(G \times G)$ always has a decomposition $Y = Y_1 \times Y_2$ satisfying the assumption of Proposition 2.3. So we get the following result.

2.5. Corollary. *Let G be an indecomposable finite group. Then the Zassenhaus conjecture is equivalent to the statement that isomorphic group bases of RG may be mapped into each other by a central automorphism. Moreover ZC-Variation i is valid for RG if, and only if, Aut-Variation i is true.*

Symmetric groups are indecomposable. Thus the two Corollaries above together with Peterson's result yield the following.

2.6. Theorem. *Let G be a direct product of symmetric groups. Then the Zassenhaus conjecture holds for RG.*

As promised we establish now some classes of finite groups for which Aut-Variation 1 is satisfied.

2.7. Theorem. *If G is a direct product of alternating groups, then Aut-Variation 1 holds.*

SKETCH OF THE PROOF: By Corollary 2.5 it suffices to prove the Theorem for A_n.

The conjugacy classes of cyclic subgroups of A_n are in bijection to the partitions of n, which correspond to even permutations of S_n. If $(k_1 \cdot 1, ..., k_n \cdot n)$ is such a partition, then the number of summands of the corresponding class sum of the conjugacy class of the cyclic subgroup generated by a permutation π with k_i i-cycles, $1 \le i \le n$, is precisely the length of the conjugacy class of π in S_n. This number is well known and given as

$$(50) \qquad \frac{n!}{k_1! \cdot 1^{k_1} \cdot k_2! \cdot 2^{k_2} \cdot ... \cdot k_n! \cdot n^{k_n}}.$$

By Theorem V.1 a normalized automorphism of RA_n permutes the class sums of conjugacy classes of cyclic subgroups. As in the case of class sums of conjugacy classes of elements (compare Remarks 1.8 this defines an action of $Aut_n(RA_n)$ on the class sums of cyclic subgroups. Moreover class sums of cyclic subgroups in the same orbit have the same number of summands, the same order of their representatives and the orbits are compatible with the power map. In order to establish Aut-Variation 1 for A_n it suffices to show that all orbits of this action have length 1.

In a first step we show that the class sums corresponding to the partitions

$$((n-4) \cdot 1, 2 \cdot 2, 0 \cdot 3, ..., 0 \cdot n) \text{ and}$$

$$((n-3) \cdot 1, 0 \cdot 2, 1 \cdot 3, 0 \cdot 4, ..., 0 \cdot n)$$

are fixed under the action of $Aut_n(A_n)$ provided $n \ge 4$ and $n \ne 6$. Expression 50 leads to the discussion of the equations

$$(n-4)! \cdot 2! \cdot 2 = k_1! \cdot k_2! \cdot 2^{k_2} \text{ and}$$

$$(n-3)! \cdot 3 = k_1! \cdot k_3! \cdot 3^{k_3}.$$

The first equation enforces $k_1 = n - 4$ and $k_2 = 2$, whileas the second one yields $k_1 = n - 3$ and $k_3 = 2$ provided $n \ne 6$. If $n = 6$, then the only non-trivial solution of the second equation is $k_1 = 0$ and $k_3 = 2$ (compare Examples 2.2).

We demonstrate this in the case of the second equation. Note that $n = k_1 + 3k_3$. This yields $(n-3)! = (n-3k_3)! \cdot k_3! \cdot 3^{k_3-1}$. Now apply the natural logarithms to both sides. If $n > 6$ and $k_3 > 1$ we get, that the left hand side is bigger than the right one. For $k_3 = 1$ we get the trivial solution. $n \le 5$ implies $k_3 = 1$. Finally for $n = 6$ we have the equation $3! = (6 - 3k_3)! \cdot k_3! \cdot 3^{k_3-1}$, which obviously gives the exceptional solution $k_3 = 2$ and $k_1 = 0$.

The cases of the alternating groups of degree less or equal 6 are now done by direct inspection (cf. Examples 2.2 for the case $n = 6$).

If $n \geq 7$, the proof proceeds by downward induction on the number of fixed points of the permutation corresponding to the partition. Note that the first step gives the start for this induction. For this part of the proof we refer to [Ki; 91, pp.93-94]. $\square$

2.8. Theorem. *Let $G = Sz(q)$ or $G = PSL(2,q)$. Then Aut-Variation 1 holds for RG.*

PROOF: For the used properties of $PSL(2, q)$ we refer to [Hu; 67], for those of $Sz(q)$ see [H-B3; 82, XI,§3 - §5]. Let K be a collection of subgroups of G . Then K is called a partition of G, if for each non-identity g of G there is precisely one element H of K such that $g \in H$.

In the case of $PSL(2, q)$

$$K = \{P^x, C^x, Z^x; x \in PSL(2, q)\}$$

form a partition of G, where P is a Sylow p-subgroup of G, C and Z are cyclic subgroups of order $(q-1)/k, (q+1)/k$ rsp. with $k = (2, q-1)$. Moreover P is isomorphic to the additive group of the field of q elements. All non-trivial elements of P are conjugate in $PGL(2, q)$. Therefore there is precisely one conjugacy class of cyclic subgroups of $PSL(2, q)$ of the same order in $PGL(2, q)$. Hence with a conjugation in $PGL(2, q)$ each normalized automorphism of $R PSL(2, q)$ may be modified to one which fixes all class sums of conjugacy classes of cyclic subgroups. So Aut-Variation 1 holds.

The Suzuki groups have a partition

$$K = \{F^x, H^x, C^x, Z^x; x \in Sz(q)\}$$

where H, C, Z are cyclic of coprime order and F is a Sylow 2-subgroup of exponent 4. All cyclic subgroups of order 2 and 4 of F are conjugate in the normalizer of F. Therefore cyclic subgroups of $Sz(q)$ of the same order are conjugate in $Sz(q)$. Hence Aut-Variation 1 holds. $\square$

§3   Brauer trees

In §2 we have seen that the study of the action of $Aut_n(RG)$ on $Cl(G)$ is sufficient to establish the Zassenhaus conjecture (or its variations) for special groups. The object of this paragraph is to demonstrate that the study of the Brauer graph, in particular the use of Brauer trees improves the method of §2.

For convenience we assume that $R = \mathbb{Z}$. Denote by $Irr(G)$ the set of the ordinary irreducible characters of G. $Aut_n(\mathbb{Z}G)$ acts as well on $Irr(G)$. In fact $\sigma \in Aut_n(\mathbb{Z}G)$ extends $\mathbb{C}$-linearly to $\mathbb{C}G$ and we identify the elements of $Irr(G)$ with the blocks of the Wedderburn decomposition of $\mathbb{C}G$.

By Proposition 1.4 and Theorem V.1 this action is compatible with the action of $Aut_n(\mathbb{Z}G)$ on $Cl(G)$. More precisely let $\alpha \in Aut_n(\mathbb{Z}G)$ then for each $\chi \in Irr(G)$ and each $C \in Cl(G)$ we have, that

$$\chi^{\alpha}(C) = \chi(C^{\alpha}).$$

By a well known result of Brauer [Hu; 67, V;13.5], if Π is a permutation group acting compatible on $Irr(G)$ and $Cl(G)$, then the number of fixed points of $\pi \in \Pi$ on $Irr(G)$ coincides with that one on $Irr(G)$. Moreover the number of orbits on $Irr(G)$ and $Cl(G)$ match up.

The automorphism group $Aut(CT(G))$ of the ordinary character table $CT(G)$ is defined as the group of all permutations π of $Irr(G)$ and $Cl(G)$ such that for each $\chi \in Irr(G)$ and each $C \in Cl(G)$ the equation

$$\chi^{\pi}(C) = \chi(C^{\pi})$$

holds. Our previous discussion shows that we have a homomorphism of $Aut_n(\mathbb{Z}G)$ into $Aut(CT(G))$ and, since the central automorphisms of $\mathbb{Z}G$ are precisely those representing the kernel of this homomorphism, we obtain the following.

3.1. Criterion. *Let G be an indecomposable group. Assume that each element of $Aut(CT(G))$ comes from a group automorphism of G. Then the Zassenhaus conjecture is valid for G.*

Note that for abelian groups $Aut(CT(G)) \cong Out(G)$. The counterexample to the Zassenhaus conjecture however shows that there are metabelian even supersoluble groups such that $Out(G)$ is a proper subgroup of $Aut(CT(G))$.

PROOF: By assumption each automorphism of the character table may be compensated by an automorphism of G. Hence Corollary 2.5 yields the result. □

3.2. Example. Let $G = M_{11}$, the Matthieu group of degree 11. Then the character table of M_{11} looks as follows.

	C_1	C_2	C_3	C_4	C_5	C_6	C_7	C_8	C_9	C_{10}
χ_1	1	1	1	1	1	1	1	1	1	1
χ_2	10	2	1	2	0	−1	0	0	−1	−1
χ_3	10	−2	1	0	0	1	$i2$	$-i2$	−1	−1
χ_4	10	−2	1	0	0	1	$-i2$	$i2$	−1	−1
χ_5	11	3	2	−1	1	0	−1	−1	0	0
χ_6	16	0	−2	0	1	0	0	0	$b11$	**
χ_7	16	0	−2	0	1	0	0	0	**	$b11$
χ_8	44	4	−1	0	−1	1	0	0	0	0
χ_9	45	−3	0	1	0	0	−1	−1	1	1
χ_{10}	55	−1	1	−1	0	−1	1	1	0	0

One sees easily that $Aut(CT(M_{11})) \cong C_2 \times C_2$. However $Out(M_{11}) = 1$. Therefore we cannot apply criterion 3.1. We remark that the character table shows that the square of each normalized automorphism of $\mathbb{Z}M_{11}$ is central and also that Aut-Variation 1, ZC-Variation 1 rsp. hold for M_{11}.

So far we used only ordinary representation theory in our study of $Aut_n(\mathbb{Z}G)$. However we can pass from $\mathbb{Z}$ via reduction mod p to the field of p-elements F_p and then by tensoring up to an algebraically closed field k of characteristic p. Hence modular representation theory enters the picture. The basic observation is the following. Each normalized automorphism α of $\mathbb{Z}G$ induces an auto equivalence on the category of its modules (sending a module M to its twisted version M^α with G acting as $\alpha(G)$). On the other hand it induces an automorphism of F_pG and hence on kG. In particular it permutes the simple kG-modules S, its projective covers P_S and the blocks of kG. Clearly the principal block will be always fixed, since the trivial simple module is fixed by any such automorphism.

3.3. Brauer graph. Fix a prime p and let $\chi \in Irr(G)$. Denote by χ_p the restriction of χ to the conjugacy classes of p-regular elements of G. Let $IBr(G)$ be the irreducible Brauer characters (*i.e.* the traces on the p-regular classes afforded by the simple kG-modules). Then

$$\chi_p = \Sigma_{\varphi \in IBr(G)} d_{\chi,\varphi} \cdot \varphi$$

for non negative integers d_φ. The integers $d_{\chi,\varphi}$ are called the decomposition numbers and they lead to the following definition of the **Brauer graph** $Br(G)$.

The vertex set of $Br(G)$ is given as $\{\chi_p ; \chi \in Irr(G)\}$. The edges are the irreducible Brauer characters. Two different vertices χ, ξ are linked by an edge if, and only if, there exists a $\varphi \in IBr(G)$ with $d_{\xi,\varphi}$ and $d_{\chi,\varphi}$ are both nonzero.

The connected components of $Br(G)$ represent precisely the blocks of kG. Moreover such a component consists of a point χ_p if, and only if, the degree of χ is divisible by the order of a Sylow p-subgroup of G and the corresponding block has defect zero. For a more detailed discussion of Brauer graphs see [Is; 76, Ch.15].

3.4. Brauer tree. Again p is a fixed prime. Well known Theorems of Brauer and Dade show that, if the defect group of a block B is cyclic, then the corresponding component of the Brauer graph is actually a tree. In particular all blocks are trees when G has cyclic Sylow p-subgroups. Such a **Brauer tree** provides more information for the module theory of the corresponding block. In contrast to the paragraph above we look now more module-theoretically than character-theoretically at these Brauer trees.

The most important information for our purpose is the following. With a given **orientation** (usually designed counterclockwise) it provides a precise combinatorial rule to calculate the composition series of the projective cover

P_S of an "edge" S (the edge is now labelled by the simple module affording the irreducible Brauer character corresponding to this edge). This rule (in particular with respect to projective resolutions often called Green's walk around the Brauer tree) is explicetely described in [Al; 86, pp.118-121], compare also the example below. For our purposes it suffices to describe its theoretical background.

Assume that the tree has at least one edge. Let S be an edge of the tree. Then by Brauer-Dade-Green $rad(P_S)/soc(P_S)$ is the direct sum of two uniserial modules (a module is called uniserial, if it has a unique composition series). These uniserial modules are physically unique and it is not forbidden that one of the uniserial modules is the zero module (the latter will be the case when one of the vertices of the edge is an end of the tree). Each such uniserial module corresponds to one of the two vertices of the edge. Its composition factors are given by the other edges which are adjacent to this vertex. In particular the intersection of their composition factors is empty.

Now let $\alpha \in Aut_n(\mathbb{Z}G)$. If α fixes a block of kG (recall that the principal block is fixed), then α permutes the simple modules S in this block, hence its projective covers, whence the corresponding uniserial modules. Note, if M is a uniserial kG-module with composition factore C_i , then $\alpha(M)$ is uniserial with composition factors $\alpha(C_i)$. Consequently α does not fix an edge S only. if the composition series of its projective cover can be a composition series of $P_{\alpha(S)}$. This gives the desired additional information as demonstrated now in the case of $G = M_{11}$.

3.5. Example. M_{11} has for $p = 5$ and $p = 11$ Sylow subgroups of order $5, 11$ rsp. Hence the Brauer graph of their principal blocks is a tree. They look as follows.

$p = 5$

$$
\begin{array}{ccccc}
 & & \circ^6 & & \\
 & & | & & \\
\circ^1 & - & \circ^8 & - & \circ^5 \\
 & & | & & \\
 & & \circ_7 & &
\end{array}
$$

$p = 11$

$$
\begin{array}{ccccccc}
 & & & & \circ^3 & & \\
 & & & & | & & \\
\circ^1 & - & \circ^2 & - & \circ^9 & - & \bullet^{6,7} \\
 & & & & | & & \\
 & & & & \circ_4 & &
\end{array}
$$

The vertices of the trees are labelled by the indices i of the ordinary irreducible characters in $CT(M_{11})$ given in example 3.2. A further result of the theory of blocks with cyclic defect groups is, that the map $\sigma : \chi \rightarrow \chi_p$ is injective with one possible exception. This exception happens in our example for $p = 11$ (not for $p = 5$). Characters with this property are usually called exceptional and $|\sigma^{-1}(\chi_p)|$ is called their multiplicity e.

We also remark that the Brauer trees of our example (as an exception) may be calculated purely from information available from the ordinary character table of M_{11}, given in Example 3.2. The methods for the calculation are described explicitly in [H-L; 90]. So the only information added to the ordinary character table is that one of the rule for the calculation of the projective covers. Looking at the character table we know already that $\alpha \in Aut_n(\mathbb{Z}G)$ induces on $Irr(M_{11})$ only possibly the non-trivial permutations (χ_3, χ_4), (χ_6, χ_7) and (χ_3, χ_4), (χ_6, χ_7). So all other irreducible characters are fixed by α. Denote by S_i the simple module labelling the edge between the vertices 8 and i in the Brauer tree for $p = 5$. Then the rule for the projective covers P_{S_6} and P_{S_7} gives the composition series

$$
\begin{array}{cc}
S_6 & S_7 \\
S_1 & S_5 \\
S_7 & S_6 \\
S_5 & S_1 \\
S_6 & S_7
\end{array}
$$

Consequently the Brauer tree for $p = 5$ shows that χ_6 and χ_7 have to be fixed. Analogously we conclude this with respect to χ_3 and χ_4 by the tree for $p = 5$. Consequently the Zassenhaus conjecture holds for M_{11}.

XI   Group Extensions

By Alexander Zimmermann

All groups in this section are assumed to be finite.

We want to study, roughly speaking, to what extend the Zassenhaus conjecture for an integral group ring of a finite group G forces the integral group ring of an abelian extension E of G to have a positive answer to the isomorphism problem.

The key idea is to look at the so called small group ring $S(A, E)$ of a group E relative to an abelian normal subgroup A. In this case the small group ring is a large quotient of the integral group ring $\mathbb{Z}E$ namely

$$S(E, A) := \mathbb{Z}E/(I(A)I(E)),$$

$I(A)$ and $I(E)$ being the corresponding augmentation ideals.

§ 1   Isomorphic Group Extensions

We introduce two categories associated with a finite group G. The objects of $\mathcal{E}_{gr}$ are short exact sequences of finite groups with abelian kernel

$$1 \longrightarrow N \longrightarrow E \longrightarrow G \longrightarrow 1.$$

When N is a $\mathbb{Z}G$–module, G operates by conjugation with a preimage in E. Morphisms are morphisms over G, i.e. commutative diagrams

$$
\begin{array}{ccccccccc}
1 & \longrightarrow & N & \longrightarrow & E & \longrightarrow & G & \longrightarrow & 1 \\
  & & \downarrow & & \downarrow & & \| & & \\
1 & \longrightarrow & N_1 & \longrightarrow & E_1 & \longrightarrow & G & \longrightarrow & 1
\end{array}
$$

The second category is $\mathcal{E}_{alg}$. Objects are short exact sequences of $\mathbb{Z}$–algebras

$$0 \longrightarrow N \longrightarrow S \longrightarrow \mathbb{Z}G \longrightarrow 0$$

for an ideal N of square zero of the ring S. Morphimsms are morphisms over $\mathbb{Z}G$, i.e. commutative diagrams analogous to the ones above.

Without proof we state:

1.1. Proposition. *The categories $\mathcal{E}_{alg}$ and $\mathcal{E}_{gr}$ are isomorphic.*

In terms of equivalence classes of short exact sequences this is nothing but the natural isomorphism

$$H^2(G, N) \simeq H^2_{alg}(\mathbb{Z}G, N).$$

Note that an isomorphism of group extensions in the above sense, i.e. an isomorphism over G, is a much stronger condition than the groups being isomorphic.

1.2. Proposition. *Given two extensions*

$$\varepsilon_G : 1 \longrightarrow N \longrightarrow E \longrightarrow G \longrightarrow 1,$$

$$\varepsilon'_G : 1 \longrightarrow N_1 \longrightarrow E_1 \longrightarrow G \longrightarrow 1$$

over G with abelian normal subgroups N, N_1 of the groups E, E_1. The above extensions are isomorphic iff there exists a $\mathbb{Z}G$-module isomorphism

$$I(N)S(N, E) \longrightarrow I(N_1)S(N_1, E_1)$$

and a ring isomorphism

$$S(N, E) \longrightarrow S(N_1, E_1)$$

such that the following diagram commutes:

$$
\begin{array}{ccccccccc}
\varepsilon_{\mathbb{Z}G} : & 0 & \longrightarrow & I(N)S(N, E) & \longrightarrow & S(N, E) & \longrightarrow & \mathbb{Z}G & \longrightarrow & 0 \\
 & & & \downarrow & & \downarrow & & \| & & \\
\varepsilon'_{\mathbb{Z}G} : & 0 & \longrightarrow & I(N_1)S(N_1, E_1) & \longrightarrow & S(N_1, E_1) & \longrightarrow & \mathbb{Z}G & \longrightarrow & 0
\end{array}
$$

1.3. Remark. We call the sequence

$$\epsilon_{alg} : 0 \longrightarrow I(N)S(N, E) \longrightarrow S(N, E) \longrightarrow \mathbb{Z}G \longrightarrow 0$$

associated to the group sequence

$$\epsilon_{gr} : 1 \longrightarrow N \longrightarrow E \longrightarrow G \longrightarrow 1.$$

The relation of being associated is functorial in the sense introduced above. [C-E; 56, R-R; 79]

PROOF OF PROPOSITION 1.2 : This Theorem is an easy consequence of Theorem 3.3 of section V. However, we recall the following constructions proving the proposition constructively:

If we start with a group sequence

$$1 \longrightarrow N \longrightarrow E \longrightarrow G \longrightarrow 1,$$

we construct the sequence of rings

$$0 \longrightarrow I_{\mathbb{Z}}(N)E \longrightarrow \mathbb{Z}E \longrightarrow \mathbb{Z}G \longrightarrow 0.$$

with ideal $I_{\mathbb{Z}}(N)E$ of square zero. From this we may construct, factoring $I(N)I(E)$, the sequence

$$0 \longrightarrow N \longrightarrow S(E, N) \longrightarrow \mathbb{Z}G \longrightarrow 0$$

by Theorem 3.3 of section V.

Conversely given such a sequence we get the sequence

$$0 \longrightarrow N \longrightarrow S(E,N) \rtimes G \longrightarrow \mathbb{Z}G \rtimes G \longrightarrow 0.$$

Forming the pullback along the diagonal embedding $\Delta : G \longrightarrow \mathbb{Z}G \rtimes G$, the resulting group sequence is again

$$1 \longrightarrow N \longrightarrow E \longrightarrow G \longrightarrow 1.$$

$\square$

1.4. Remark. The proof shows that a group extension with abelian kernel ϵ_{gr} splits if and only if the associated extension ϵ_{alg} splits.

§2   A first Application to the Isomorphism Problem

In 1968 Whitcomb [Wh; 68] and Jackson [Ja; 69] [29] proved that the isomorphism problem for group rings $\mathbb{Z}G$ of metabelian groups G (i.e. groups with derived length 2) has a positive answer. This result was regarded as the best concerning the isomorphism problem until K. W. Roggenkamp and L.L. Scott achieved the breakthrough during the last decade [R-S; 87 1]. The key idea is to apply Proposition 1.2. We first want to present Whitcomb's theorem here.

2.1. Theorem. *Let E be a finite group with abelian commutator group N. Then the group ring $\mathbb{Z}E$ determines augmented group bases [30] up to isomorphism.*

2.2. Remark.

(a) There is no loss of generality to assume that group bases are augmented. In fact if ε is the augmentation of $\mathbb{Z}G$. Then ε is a ring homomorphism and for every group basis H of $\mathbb{Z}G$ there is an isomorphic augmented group basis
$$\tilde{H} := \{\varepsilon(h)^{-1} \cdot h | h \in H\}.$$

(b) This result holds more generally for $\mathbb{Z}$ replaced by a Dedekind domain R of characteristic zero such that no rational prime divisor of the order of E is invertible in R.

PROOF OF THEOREM 2.1: Let $\mathbb{Z}E = \mathbb{Z}E_1$ as augmented algebras. Then there exists N_1 normal and abelian in E_1 such that

$$I_{\mathbb{Z}}(N)E = I_{\mathbb{Z}}(N_1)E_1$$

---

[29]   The paper of Jackson has not obtained the credit it devices. Maybe the reason is that there are several gaps in the argument; however, these can be fixed [Ro; 81]. The paper contains a wealth of detailed information on the structure of metabelian group rings and on the group of units of such a group ring.

[30]   these are group bases of augmentation 1

by the normal subgroup correspondence IV.1. With a suitable identification we put $G_1 = E_1/N_1$ such that $\mathbb{Z}E/N = \mathbb{Z}G_1$. Therefore since E/N is abelian we apply Theorem IV.1 to conclude $E/N = G_1$ in $\mathbb{Z}G$. Therefore the preimages E and E_1 of G and G_1 in $S_{\mathbb{Z}}(N, E) = S_{\mathbb{Z}}(N_1, E_1)$ coincide, and the theorem is proved.   □

A very similar argument is used to prove the same theorem in section V.

2.3. Remark. Observe that we only need that group bases of $\mathbb{Z}E/E'$ are subsets of $E/E' \cup -E/E'$ and that E' is abelian.

Before we come to the next generalization we have to recall a criterion when two split group extensions are isomorphic:

2.4. Lemma. *Let E and E_1 be finite groups with abelian, normal, complemented subgroups N and N_1 resp. and put $G := E/N$ and $G_1 := E_1/N_1$. If there is an isomorphism $\alpha : G \longrightarrow G_1$ such that the twisted left $\mathbb{Z}G$-module $^{\alpha}N_1$ is isomorphic to N as $\mathbb{Z}G$-module* [31], *then the group extensions*

$$1 \longrightarrow N \longrightarrow E \longrightarrow G \longrightarrow 1$$

and

$$1 \longrightarrow N_1 \longrightarrow E_1 \longrightarrow G_1 \longrightarrow 1$$

are isomorphic via α.

PROOF OF LEMMA 2.4: We form the following commutative diagram with exact rows via pullback and pushout:

$$
\begin{array}{ccccccccc}
1 & \longrightarrow & N_1 & \longrightarrow & E_1 & \longrightarrow & G_1 & \longrightarrow & 1 \\
 & & \| & & \uparrow & & \uparrow \alpha & & \\
1 & \longrightarrow & ^{\alpha}N_1 & \longrightarrow & ^{\alpha}N_1 \rtimes G & \longrightarrow & G & \longrightarrow & 1 \\
 & & \downarrow \sigma & & \downarrow & & \| & & \\
1 & \longrightarrow & N & \longrightarrow & E & \longrightarrow & G & \longrightarrow & 1
\end{array}
$$

Therefore the result follows since σ is G–linear and hence the map

$$N \rtimes G \overset{(\sigma^{-1}, \alpha)}{\longrightarrow} N_1 \rtimes G_1$$

is an isomorphism.   □

An application of this lemma is the following:

2.5. Theorem. *[Zi; 90] Let N be an abelian finite group and let G be a group of automorphisms of N with $(|G|, |N|) = 1$ such that augmented group bases of $\mathbb{Z}G$ are conjugate in $\mathbb{Q}G$. Then $\mathbb{Z}(N \rtimes G)$ determines group bases up to isomorphism.*

---

31   with $^{\alpha}N_1$ is N_1 as $\mathbb{Z}$-module but $g \in G$ acts as $\alpha(g)$

PROOF OF 2.5: Let $\mathbb{Z}E = \mathbb{Z}E_1$ as augmented $\mathbb{Z}$–algebras. From Theorem V.1 in section V we know that for $N \trianglelefteq E$ there exists $N_1 \trianglelefteq E_1$ such that $I(N)E$ is equal to $I(N_1)E_1$. Let $G_1 = E_1/N_1$ by a suitable identification such that $\mathbb{Z}G = \mathbb{Z}G_1$. By assumption we have a central automorphism α of $\mathbb{Z}G$ such that

$$\alpha(G) = G_1.$$

Let R be the semilocalization of $\mathbb{Z}$ at the primes dividing $|N|$. From Lemma 2.13 of section VII we see that $Picent(RG)$ is trivial. On the other hand $Outcent(RG) \leq Picent(RG)$ and so we see that α is in fact conjugation by a unit a of RG.

We define

$$S_R(N, E) := R \otimes_{\mathbb{Z}} S(N, E).$$

Since

$$(I_R(N)S_R(N, E))^2 = 0,$$

the ideal $I_R(N)S_R(N, E)$ lies in the radical of $S_R(N, E)$, we apply Nakayama's lemma to observe that the unit a has a preimage b in $S_R(N, E)$. However conjugation by b induces an automorphism of $I_R(N)S_R(N, E)$– it is a twosided ideal of $S_R(N, E)$— we see that the preimage E_R must be conjugate by b to the preimage $(E_1)_R$ of E_1. Therefore

$$N = R \otimes_{\mathbb{Z}} N \simeq R \otimes_{\mathbb{Z}} N_1 = N_1.$$

Hence the group extensions

$$1 \longrightarrow R \otimes_{\mathbb{Z}} N \longrightarrow E_R \longrightarrow G \longrightarrow 1$$

and

$$1 \longrightarrow R \otimes_{\mathbb{Z}} N_1 \longrightarrow (E_1)_R \longrightarrow G \longrightarrow 1$$

are isomorphic as well as the extensions

$$0 \longrightarrow I_R(N)S_R(N, E) \longrightarrow S_R(N, E) \longrightarrow RG \longrightarrow 0$$

and

$$0 \longrightarrow I_R(N_1)S_R(N_1, E_1) \longrightarrow S_R(N_1, E_1) \longrightarrow RG_1 \longrightarrow 0.$$

So we are done since therefore $E_R = E$ and $(E_1)_R = E_1$. $\square$

2.6. Corollary. *Let*

$$1 \longrightarrow N_p \times N_q \longrightarrow E \longrightarrow G \longrightarrow 1$$

be an exact sequence of finite p-solvable groups such that $N = N_p \times N_q$ is abelian and N_p is the p-Sylow subgroup of N, $O_{p'}(G) = 1$ and $(|N_q|, |G|) = 1$. Then the isomorphism problem for $\mathbb{Z}E$ has a positive answer.

PROOF: $O_{p'}(E/N_q) = 1$ and therefore we can apply the above theorem, if we choose N_p instead of N.   □

§ 3   Using subgroup rigidity

We have seen in the previous section that the discussion of automorphisms is essential for questions concerning the isomorphism problem. Now we will use the technique of invertible bimodules to get information about automorphisms of integral group rings of certain direct products of groups satisfying the strong isomorphism theorem stated above (Theorem VI.2). The method shown below is essentially due to L.L. Scott [Sc; 85], developed for proving the isomorphism theorem for finite nilpotent on abelian groups (cf. [R-S; 86]) and was generalized in [Zi; 90]. Though we have a shorter proof of the theorem below and even can extend the statement, we want to give this proof here.

The theorem we want to prove states as follows:

3.1. Theorem. *Let E be a finite solvable group with a normal p–subgroup N and complement G satisfying the Zassenhaus conjecture and which is a direct product $G = G_1 \times G_2$ of groups of relatively prime order and $O_{p'}(G) = G_2$ and furthermore assume $[N, G_1] = 1$. Let α be a normalized automorphism of $\mathbb{Z}E$ stabilizing $I(N)E$ as set, then the induced automorphism $\bar{\alpha}$ decomposes into $\bar{\alpha} = \gamma(\prod_i \sigma_i)\rho$ for an inner automorphism γ of $\hat{\mathbb{Z}}_p G$, central group automorphisms σ_i of G_1, one for each $\hat{\mathbb{Z}}_p G$–block, and a group automorphism ρ of G.*

Furthermore, if G_1 is a p–group, then the σ_i differ on the image of the principal block of E only by inner automorphisms of G_1.

If G is not necessarily a p–group then α induces an automorphism on $\mathbb{F}G$, the group ring over the field of p elements, which has the property that the σ_i differ on the image of the principal block of E only by an inner automorphism of $\mathbb{F}G_1$.

PROOF: We start with a finite solvable group G, which splits into a direct product $G = G_1 \times G_2$ of groups of relative prime order such that $O_{p'}(G) = G_2$ and G satisfies the Zassenhaus conjecture (cf. section IX). Let N be a finite complemented normal p–subgroup of the finite solvable group E with $E/N = G$ and $[N, G_1] = 1$ viewing G as a subgroup of E. Let furthermore α be an augmented automorphism of $\mathbb{Z}E$ stabilizing $I_{\mathbb{Z}}(N)E$ as set and thus inducing an automorphism $\bar{\alpha}$ of $\mathbb{Z}G$. Since G satisfies the Zassenhaus conjecture $\bar{\alpha} = \beta\rho$ for a central automorphism β of $\mathbb{Z}G$ and a group automorphism ρ of G. From XIII.3 we get that

$$\hat{\mathbb{Z}}_p G = \hat{\mathbb{Z}}_p G_1 \otimes_{\hat{\mathbb{Z}}_p} \prod_{i=1}^{t} (R_i)_{n_i} = \prod_i (R_i G_1)_{n_i}$$

for finite unramified extensions R_i of $\hat{\mathbb{Z}}_p$. Since $O_{p'}(G_1) = 1$ this is the decomposition of $\hat{\mathbb{Z}}_pG$ as algebra (see [H-B2; 82, VII;13.1]). The automorphism β is central and so it decomposes to a product of automorphisms β_i one for each factor.

Since *Picent* and *Outcent* are additive functors we get that

$$
\begin{aligned}
Outcent(\hat{\mathbb{Z}}_pG) &= \prod_{i=1}^{t} Outcent(R_iG_1)_{n_i} \\
&= \prod_i Picent(R_iG_1)_{n_i} \\
&= \prod_i Picent(R_iG_1) \\
&= \prod_i Outcent(R_iG_1)
\end{aligned}
$$

using the discussion in section VII. Since β is an integral automorphism we may apply Theorem VI.1 of section VI to conclude that $\beta_i = \gamma_i\sigma_i$ for an inner automorphism γ_i of $(R_iG_1)_{n_i}$ and central group automorphisms σ_i of G_1. (Here a group automorphism is called central, if it induces a central automorphism of the corresponding integral group ring.)

For the time being we assume the following:

3.2. Assumption. *For the simple $\mathbb{F}_pG_2$ modules S and S' with $S \not\simeq S'$ the group $Ext^1_{\mathbb{F}E}(S', S)$ is non trivial.*

From the decomposition of $\hat{\mathbb{Z}}_pG$ into indecomposable factors we know that the central idempotents of $\hat{\mathbb{Z}}_pG$ lie in $\hat{\mathbb{Z}}_pG_2$ rather than in $\hat{\mathbb{Z}}_pG$. Furthermore the two modules S and S' lie in different ring direct factors of $\hat{\mathbb{Z}}_pG$ and so $Ext_{\mathbb{F}G}(S', S) = 0$.

Using the assumption we get a non split exact sequence

$$0 \longrightarrow S \longrightarrow X \longrightarrow S' \longrightarrow 0$$

of $\mathbb{F}_pE$ modules. Since N has a complement the group G can be found in E and this special group is fixed throughout. When restricting the above sequence to G it splits obviously, since the extension group as G–modules vanishes. Since N is a p–group, the ideal $I_{\mathbb{F}}(N)E$ lies in the radical of $\mathbb{F}E$. Therefore

$$I_{\mathbb{F}}(N)E \cdot X \le (rad\mathbb{F}E)X \le rad_{\mathbb{F}E}X \le ker(X \longrightarrow S') = S$$

and since S is simple we get that $I(N)E \cdot X = S$ or $I(N)X = 0$. But in the second case N would operate trivial on X and therefore the module structure on X is induced by the module structure of G on X and so the sequence would split, a contradiction.

Since G_1 is an epimorphic image of E, the ring $\mathbb{F}G_1$ is a $\mathbb{F}E$–left–module by letting E act via $E \longrightarrow G_1$. From the right we let E act trivially on

X and thus giving a $I\!\!F E-$ bimodule structure on X and on $I\!\!F G_1$ and hence also on $I\!\!F G_1 \otimes_{I\!\!F} X$ with E acting diagonally. Since N acts trivial on $I\!\!F G_1$ we see that for all $n \in N$

$$(n-1)(m \otimes x) = (nm \otimes nx) - (m \otimes x) = m \otimes (n-1)x$$

for all $x \in X, m \in I\!\!F G$. Therefore

$$I(N)E \cdot (I\!\!F G_1 \otimes X) = I\!\!F G_1 \otimes I(N) \cdot X = I\!\!F G_1 \otimes S.$$

As usual (see section VII) we write $_\alpha M_1$ for the module M as abelian group but the action of E being twisted by α from the left. Then we define a map ψ'_n for all $n \in N$ via

$$\psi'_n : \; _\alpha(I\!\!F G_1 \otimes X)_1 \longrightarrow \; _\alpha(I\!\!F G_1 \otimes X)_1$$

which is multiplication by $n-1$ from the left. Since $[N, G_1] = 1$, the maps ψ'_n are $I\!\!F G_1$ linear. The above observations yields

$$Im \; \psi'_n \subseteq \; _\alpha(I\!\!F G_1 \otimes X)_1 \subseteq ker \; \psi'_n$$

inducing a bimodule homomorphism

$$\psi_n : \; _\alpha(I\!\!F G_1 \otimes S')_1 \longrightarrow \; _\alpha(I\!\!F G_1 \otimes S)_1 \, ,$$

which gives a homomorphism

$$\psi_n : \; _{\sigma_{j'}\rho}(I\!\!F G_1 \otimes S')_1 \longrightarrow \; _{\sigma_{i'}\rho}(I\!\!F G_1 \otimes S)_1,$$

since S lies in the ith component of $\hat{Z\!\!\!Z}_p G$ and S' in the jth, observing that inner automorphisms do not change the isomorphism class of bimodules and the block dependence of S determines that of the tensor product. Here we denote the permutation of the blocks by ρ, which is not central, by a dash. But as G_1–modules S and S' are both trivial. If we vary n in N the set $\{n-1\}$ generates $I(N)$ as vector space and so the images $Im\psi_n$ generate additively $_\alpha(I\!\!F G_1 \otimes S)_1$. Therefore for large enough k we get an epimorphism of $I\!\!F G_1$–bimodules

$$\Psi : (\; _{\sigma_{j'}\rho}(I\!\!F G_1)_1)^k \longrightarrow \; _{\sigma_{i'}\rho}(I\!\!F G_1)_1.$$

The next lemma is due to Steffen König:

3.3. Lemma. *Let A be a finite dimensional $I\!\!F$–algebra and $N_1, .., N_k$ be indecomposable A–left–modules of equal dimension d with $k \geq 2^d + 1$. Let furthermore*

$$\phi_i := (\phi_i^1, .., \phi_i^n) \in Hom_A(N_i^n, N_{i+1})$$

for a fixed n be surjective homomorphisms such that some ϕ_i^j is monic if and only if ϕ_i^1 is monic for all j and all i.

Then some $\phi_1^{j_0}$ is an isomorphism.

PROOF: We define

$$\psi_i \in Hom_A(N_i^{(n^{k-i+1})}, N_{i+1}^{(n^{k-i})})$$

in the obvious way as $\psi_i := \prod_{l=1}^{n^{k-i}} \phi_i$ and see that all ψ_i are epic. Now $\Psi :=$ $\psi_1 \psi_2 ... \psi_{k-1} : N_1^{n^k} \longrightarrow N_k$ is also epic. If we restrict Ψ to its lth component of the various copies of N_1 we see that this mapping is equal to

$$\phi_1^{j_1(l)} \phi_2^{j_2(l)} ... \phi_{k-1}^{j_{k-1}(l)}$$

for some natural numbers $j_1(l), j_2(l), .., j_{k-1}(l)$. Since the sum of all the images of the restricted mapping Ψ is equal to the image of Ψ, one of our restriction is non zero.

Now we apply the Lemma of Harada and Sai ([Ben; 84, 2.31.9]) which states that whenever a composite of 2^d mappings between indecomposable modules of composition length at most d is not zero then one of these mapping is an isomorphism. We will not prove this Lemma here and mention just that it is proved by induction and uses that the composition length of the image of the composite of $2^m - 1$ maps has composition length $d - m$.   □

In our situation $A := \mathbb{F}G_1 \otimes_\mathbb{F} \mathbb{F}G_1^{op}$, n is our k and the modules are $(\sigma_{j'} \rho)^{-1}(\sigma_{i'} \rho))^l (\mathbb{F}G_1)_1$ for suitable numbers l, the mappings then are obvious. Since $O_{p'}(G_1) = 1$ and G_1 is solvable we have that these modules are indecomposable. Therefore $\sigma_{i'}$ and $\sigma_{j'}$ differ only by an inner automorphism of $\mathbb{F}G_1$.

If G_1 is a p-group, then the conjugating element normalizes G_1 in $\mathbb{F}G_1$ and with Coleman's result in chapter II § 2.1 it is a product of an element of $\mathbb{F}G_1$ centralizing G_1 and an element of G_1. Therefore the inner automorphism above is already inner in G_1.

We now just have to verify our assumption: The most elementary characterization of blocks is that two simple modules S and T belong to the same block if and only if there is a sequence of simple modules $S = S_1, S_2, .., S_t = T$ such that for all i either $Ext^1(S_i, S_{i+1}) \neq 0$ or $Ext^1(S_{i+1}, S_i) \neq 0$ ([C-R; 62, pp. 378–380]). If we restrict our attention to the image of the principal block of $\mathbb{F}E$ (that is the block, the trivial module belongs to) then the twist by ρ, as above indicated by a dash, does not matter, since the principal block is just $\mathbb{F}E/O_{p'}(E)$ and this is fixed under α as one sees from the normal subgroup correspondence between E and $\alpha(E)$.

So we proved the theorem.   □

This theorem is the main tool for proving the isomorphism theorem for integral group rings of finite nilpotent by abelian groups:

3.4. Theorem. *([R-S; 86, Sc; 85]) Let E be a finite group with abelian normal subgroup N such that $E/N =: G$ is nilpotent, then the isomorphism problem for $\mathbb{Z}E$ has a positive answer.*

The theorem is valid even for a general Dedekind domain of characteristic zero instead of $\mathbb{Z}$ such that no prime divisors of $|E|$ are invertible in R.

PROOF OF THEOREM 3.4: Let N be the normal subgroup of E of smallest order such that E/N is nilpotent. This subgroup is unique, it is equal to $\bigcap_i K_i(E)$ with $K_0(E) = E$ and $K_i(E) = [K_{i-1}(E), E]$ ([Hu; 67, III (2.5)]). Our assumptions assure that N is abelian.

Let $\mathbb{Z}E = \mathbb{Z}E_1$ as augmented algebras. By Theorem IV.1 we get a normal subgroup M of E_1 with $I(N)E = I(M)E_1$. Now M is abelian by Theorem 3.3 of section V. Let N_p be the p–Sylow–subgroup of N and M_p that of M respectively. The analogous notation is used for G and $G_1 := E_1/M$. We define

$$A := \prod_p [N_p, P].$$

Here P' is the complement of P, the p–Sylow subgroup of G. For A there is a normal subgroup B of E_1 underneath M such that $I(A)E = I(B)E_1$. Let $\overline{E}_1 \simeq E_1/B$ be the group such that

$$\mathbb{Z}\overline{E} = \mathbb{Z}\overline{E}_1.$$

Furthermore let $\overline{N} := N/A$ and $\overline{M} \simeq M/B$ naturally in $\overline{E}_1$. Then G is the largest nilpotent quotient of $\overline{E}$ as well because otherwise a smaller normal subgroup $\overline{N}'$ of $\overline{E}$ with nilpotent quotient would have as preimage in E also a smaller normal subgroup N' of E with nilpotent quotient.

Therefore every augmented automorphism of $\mathbb{Z}\overline{E}$ preserves $I(\overline{N})\overline{E}$. We claim that the isomorphism problem is true for $\mathbb{Z}\overline{E}$. We will see that N has a complement in E.

In fact $I(P')N_p = N_p$ because otherwise $\prod_p I(P')N_p$ would be a smaller normal subgroup with nilpotent quotient. This is checked for each prime p because if N_p is normal in $E/N_{p'}$ with nilpotent quotient G. Then we form the pullback

$$\begin{array}{ccc} \prod E/N_{p'} & \longrightarrow & \prod_p G \\ & & \uparrow \Delta \\ & & G \end{array},$$

when Δ is the diagonal. Then also $N = \prod_p N_p$ is normal in E with nilpotent quotient. For the details of this kind of arguments we refer to [R-R; 79, 2.17]. Since $I(P')N_p = N_p$, the module N_p does not belong to the principal $\hat{\mathbb{Z}}_p G$ block. Therefore

$$H^2(G, N_p) = Ext^2_{\mathbb{Z}G}(\mathbb{Z}, N_p) = Ext^2_{\hat{\mathbb{Z}}_p G}(\hat{\mathbb{Z}}_p, N_p) = 0$$

and N must have a complement in E.

We proceed by showing that the isomorphism problem has a positive answer for $\mathbb{Z}\overline{E}$.

For the moment we omit the bar over the symbols. Let $N_{p'}$ be the complement to N_p in N. Analogous notations are used for M and for G_1. Since the Zassenhaus conjecture is true for nilpotent groups, we have a central automorphism α of $\mathbb{Z}G$ with $\alpha(G) = G_1$. As in the preceeding paragraph, in $\hat{\mathbb{Z}}_p G$ we have a decomposition $\alpha = \gamma(p) \prod_i \sigma_i(p)$ for inner automorphisms $\gamma(p)$ and central group automorphisms $\sigma_i(p)$ of P. Then

$$^\alpha M_p = \prod_i {}^{\gamma \sigma_i(p)} e_i M_p \simeq \prod_i {}^{\sigma_i(p)} e_i M_p \simeq M_p \simeq N_p$$

as $\mathbb{Z}G$–modules, since P operates trivially on N_p. Here e_i are the block idempotents of $\hat{\mathbb{Z}}_p G$. Our criterion now tells us that $E \simeq E_1$.

We now use again the bar over the symbols. Since $\mathbb{Z}\overline{E} = \mathbb{Z}\overline{E}_1$, we have an isomorphism $\alpha : \overline{E} \longrightarrow \overline{E}_1$. Since $\overline{N}$ is minimal among the normal subgroups of $\overline{E}$ with nilpotent quotient, α fixes $I(\overline{N})\overline{E}$ as set. Otherwise $\alpha(\overline{N})$ would be not the smallest normal subgroup of $\alpha(\overline{E})$ with nilpotent quotient, a contradiction. Now α induces an automorphism on $\mathbb{Z}G$ also denoted by α with $\alpha(G) = G_1$. Since N is abelian, α also fixes $I(\overline{N}_{p'})\overline{E}$, where in turn the induced automorphism on $\mathbb{Z}\overline{E}/\overline{N}_{p'}$ also fixes $I(\overline{N}_p)\overline{E}/\overline{N}_{p'}$. Thus we may assume for the moment that N is a p–group. Since $I(P)N_p = [N_p, P]$ is the radical of N_p as $\hat{\mathbb{Z}}_p P$–module, the quotient $\overline{N}_p$ belongs to exactly the same $\hat{\mathbb{Z}}_p G$–blocks as N_p does by Nakayama's lemma.

We extend the automorphism α_p linearly to $\hat{\mathbb{Z}}_p G$ also denoted by α_p. On the image of the principal $\hat{\mathbb{Z}}_p \overline{E}$–block we have that in the decomposition of α_p in Theorem 3.1 the maps $\sigma_i(p)$ coincide up to inner automorphisms of P as long as i is an index corresponding to the principal block of $\hat{\mathbb{Z}}_p \overline{E}$. We define

$$\sigma := \prod_p \sigma_1(p).$$

Since $\sigma_1(q)$ is a central automorphism of $\hat{\mathbb{Z}}_p G/P$ as long as $p \neq q$, it is inner in $\hat{\mathbb{Z}}_p G$. Therefore σ differs from $\sigma_1(p)$ by an inner automorphism. Since

$$[\overline{N}_p, O_{p'}(\overline{E})] \leq \overline{N}_p \cap O_{p'}(\overline{E}) = 1,$$

the module N_p belongs to the principal $\hat{\mathbb{Z}}_p \overline{E}$–block.

$$
\begin{aligned}
^\alpha M_p &= \prod_i {}^{\gamma(p)\sigma_i(p)\rho} e_{i'} M_p \\
&\simeq \prod_i {}^{\sigma_i(p)\rho} e_{i'} M_p \\
&\simeq \prod_i {}^{\sigma\rho} e_{i'} M_p \\
&\simeq \prod_i {}^{\sigma\rho} e_{i'} N_p
\end{aligned}
$$

with $\rho^{-1}(e_i) = e_{i'}$. The last isomorphism holds, since in the small group ring, N and M are isomorphic. Our criterion tells us now that E and E_1 are isomorphic.

$\square$

3.5. Remark. We used the fact, that G is nilpotent at various places. However a slight change of the proof gives a positive answer to the isomorphism problem for finite groups having an abelian normal subgroup N with quotient being a direct product of a nilpotent group G and a group H such that $(|H|, |G|) = 1$ both satisfying the Zassenhaus conjecture such that the normal subgroup N and H are of relatively prime order. We only have to prove the analogue of the fact that there is a unique normal subgroup N of smallest order such that the quotient is of the required form. This and the proof of Theorem 3.1 in the above generality is worked out in [Zi; 90]. Theorem 3.1 is proved in [R-S; 86] and [Sc; 85] under the hypotheses that G is nilpotent and the above proof follows closely these arguments.

§4   The main result

The most far reaching result in this direction is a consequence of [Sc; 87, Sc; 90] and will be published elsewhere. However we will state the theorem and give a short idea of how the proof works.

4.1. Theorem. *Let $\mathcal{P}$ be a finite set of n rational primes and let*

$$G = \prod_{p \in \mathcal{P}} G_p \times H \ , \ \text{with } G_p \ p\text{-solvable and } O_{p'}(G_p) = 1,$$

be a product of $n + 1$ groups of relative prime order such that H satisfies the Zassenhaus conjecture, i.e. augmented group bases in $\mathbb{Z}H$ are conjugate in $\mathbb{Q}H$. Furthermore, let E be a finite group and let N be an abelian normal subgroup with $E/N \simeq G$. For P_p being a Sylow p–subgroup of G_p let

$$(|N|, \frac{|G|}{\prod_{p \in \mathcal{P}} |P_p|}) = 1 \, .$$

Then the isomorphism problem for $\mathbb{Z}E$ has a positive answer, i.e. $\mathbb{Z}E \simeq \mathbb{Z}E_1$ implies $E \simeq E_1$.

We improved the method of the last paragraph in using Theorem VI.1. This and an analogue to the well known statement that there is a unique minimal normal subgroup of a group with minimal quotient are the main tools in the proof of Theorem 4.1. The procedure is to prove first that the theorem is true, if in addition N is a semi simple $\mathbb{Z}G$–module and afterwards factor out in the general case the radical of N as $\mathbb{Z}G$–module. The automorphism of the first step gives an automorphism of $\mathbb{Z}G$, which can be lifted to the small group ring if one uses a suitable generalization of the result in the last paragraph. This in turn is possible if one uses Theorem VI.1. The rest of the argument is the same as above.

To give an idea of the groups we cover we specialize to $|\mathcal{P}| = 1$ and get the

4.2. Corollary. *Let G be a finite solvable group with normal Hall π-subgroup H and complement K for H in G. Assume furthermore that G has an abelian normal subgroup N in H of p-power order, such that $K \trianglelefteq G/N$ and every non trivial normal subgroup of H/N has an order divisible by p. Then a positive answer to the Zassenhaus conjecture for $\mathbb{Z}K$ implies a positive answer to the isomorphism problem for $\mathbb{Z}G$.*

XII Class sums of p-elements

by W.Kimmerle

§ 1 Groups with nilpotent commutator subgroup

The object of this section is to establish further classes of groups which satisfy the Zassenhaus Conjecture Variation 2, cf. section X.

Before we start with this we discuss the previous sections from a group-theoretically point of view.

Recall that the Zassenhaus conjecture is valid for the class N of nilpotent groups [R-S; 87 1], see also section VI. Towards the way to soluble groups we have seen in section IX that this conjecture does not hold for metabelian groups (the class of metabelian groups we denote by AA). In section XI it was explained that groups G with abelian normal subgroup A and G/A nilpotent still have a positive solution for the isomorphism problem (we denote this class of groups by AN).

We summarize the development of the isomorphism problem (IS) and the Zassenhaus conjecture (ZC) in the following picture.

$$
\begin{array}{ccc}
A & \subset & AA \\[4pt]
\text{IS} \wedge \text{ZC [Hig; 39]} & & \text{IS [Wh; 68]} \\
\text{covered by IV.1} & & \neg \text{ ZC see IX §1} \\[8pt]
\cap & & \cap \\[8pt]
N & \subset & AN \\[4pt]
\text{IS} \wedge \text{ZC [R-S; 87 1]} & & \text{IS [R-S; 86][R-Z; 90]} \\
\text{covered by VI and X} & & \neg \text{ ZC see IX §1}
\end{array}
$$

It is consequently natural to consider the class of groups which have a nilpotent commutator subgroup (we denote this class by NA). Supersoluble groups are an interesting class of groups between nilpotent and soluble groups. In terms of chief factors they are precisely those soluble groups with simple chief factors. Moreover the class of supersoluble groups is contained in NA.

That ZC-Variation 2 holds for supersoluble groups is shown in [Ki; 91, 5.20]. In this paragraph we shall extend this result to groups with nilpotent commutator subgroup. Note that the counterexample to the Zassenhaus conjecture is not only metabelian it is also a supersoluble group.

Supersoluble groups are precisely those soluble groups whose chief factors are simple. It seems to be a natural question whether the isomorphism problem may be proved in the situation for a general group with simple chief factors.

The special case when all chief factors of a finite group are isomorphic to one and the same simple group should be regarded as the general analogue of a p-group. With respect to the isomorphism problem for such groups we have indeed a positive answer for the isomorphism problem combining the nilpotent group result of Roggenkamp and Scott with Theorem 2.2 of section X and the well known Schreier Conjecture. Schreiers conjecture holds by the classification of the finite simple groups and it implies that a group with all chief factors being simple non-abelian is semi simple. Hence the isomorphism problem (and "sometimes the Zassenhaus conjecture", cf. section IX) holds for such a group, cf. [K-L-S-T; 90, 2.8].

1.1. Theorem. *Let G be in the class NA. Then ZC-Variation 2 holds for G.*

We explain first the **strategy** for the proof of Theorem 1.1. Let G be a finite group and let M and N be normal subgroups of G with $M \cap N = 1$. Then we can write G as a pullback of the following form.

$$\begin{array}{ccc} G & \xrightarrow{\alpha} & G/M \\ \downarrow{\gamma} & & \downarrow{\beta} \\ G/N & \xrightarrow{\delta} & G/(M \cdot N) \end{array}$$

where β, α, γ and δ are the reduction maps. We assume further that M and N (and therefore $M \cdot N$) are characteristic. We shall mainly apply this diagram with $M = O_{p'}(G)$ and $N = O_p(G)$.

The strategy is now to assume that G/N and G/M are somehow well behaved. The question is then, which of the good properties carry over to G.

The group homomorphisms α, β, γ and δ induce, since M and N are characteristic, ring automorphisms denoted for convenience with the same symbols, making the following diagram (which is **not a pullback diagram of rings**) commutative.

$$\begin{array}{ccc} \mathbb{Z}G & \xrightarrow{\alpha} & \mathbb{Z}G/M \\ \downarrow{\gamma} & & \downarrow{\beta} \\ \mathbb{Z}G/N & \xrightarrow{\delta} & \mathbb{Z}G/(M \cap N) \end{array}$$

Now let $\tau \in Aut_n(\mathbb{Z}G)$. Assume that τ fixes

$$I(M) \cdot \mathbb{Z}G, \ I(N) \cdot \mathbb{Z}G \text{ and } I(M \cdot N) \cdot \mathbb{Z}G.$$

Then τ induces automorphisms

$$\tau_N \in Aut_n(\mathbb{Z}G/N), \tau_M \in Aut_n(\mathbb{Z}G/M) \text{ and } \tau_{MN} \in Aut_n(\mathbb{Z}G/(M \cdot N)).$$

$(i.e. \tau_N(\gamma(x)) = \gamma(\tau(x)))$ Assume that $G/(M \cdot N)$ is abelian. Then τ_{MN} is just given by a group automorphism α_{MN} of $G/M \cdot N$.

Moreover, since τ_{MN} comes from τ_M and τ_N resp., it is reasonable to **assume that τ_M and τ_N can be written in the form**

$$\tau_M = \alpha_M \cdot \sigma_M \text{ and } \tau_N = \alpha_N \cdot \sigma_N \,,$$

accordingly to the Aut-Variation 2.

Since σ_M and σ_N fix the class sums of conjugacy classes of elements of prime power order, it follows that α_{MN} comes from α_M and α_N resp.

1.2. Claim. α_M and α_N may be lifted to a group automorphism α of G.

PROOF: Since G is a pullback, it follows that G is isomorphic to the subgroup

$$\{(x,y); x \in G/M, y \in G/N, \beta(x) = \delta(x)\}$$

of $G/M \times G/N$. Clearly $\alpha_M \times \alpha_N$ yields an automorphism of $G/M \times G/N$. We show that this yields an automorphism α_1 of the subgroup isomorphic to G (for the sake of simplicity we identify in the sequel this subgroup with G).

It suffices to show that $\beta(\alpha_M(x)) = \delta(\alpha_N(y))$. The previous constructions show that $\beta(\alpha_M(x)) = \alpha_{MN}(\beta(x))$ and $\delta(\alpha_N(y)) = \alpha_{MN}(\delta(y))$. By assumption we know that $\beta(x) = \delta(y)$ and the desired equality follows.   □

Therefore we can modify the automorphism τ by α_1^{-1} . The automorphism $\tau_1 = \alpha_1^{-1} \cdot \tau$ induces then on $\mathbb{Z}G/N$ the automorphism σ_N and on $\mathbb{Z}G/M$ the automorphism σ_M.

1.3. Lemma. Let $1 \to K \to G \xrightarrow{\pi} G/K \to 1$ be an exact sequence of groups. Assume that the prime p does not divide $|K|$ and let x and y be p-elements (i.e. x is of order p^n for some $n \in \mathbb{N}$) in G. Then x is conjugate to y in G if, and only if, $\pi(x)$ and $\pi(y)$ are conjugate in G/K.

PROOF: One direction is obvious. Assume that $\pi(x)$ and $\pi(y)$ are conjugate in G/K by $\pi(z)$. Consider the exact sequence

$$1 \to K \to \pi^{-1}(< \pi(x) >) \xrightarrow{\pi} < \pi(x) > \to 1.$$

By the previous y is conjugate in G to an element $y_1 \in \pi^{-1}(< \pi(x) >)$ with $\pi(y_1) = \pi(x)$.

Because $(|\pi(x)|, |K|) = 1$ we know by the Schur-Zassenhaus Theorem that the sequence splits and that all complements to K in $\pi^{-1}(< \pi(x) >)$ are conjugate by elements of K. Thus there exists $k \in K$ with $< y_1 >^k = < x >$. Also $\pi(y_1^k) = \pi(x)$. However, $\pi_{|<x>}$ is an isomorphism. Hence we obtain that $y_1^k = x$ and putting things together it follows that y is conjugate to x.   □

1.4. Corollary. Let $1 \to K \to G \to G/K \to 1$ be an exact sequence of groups. Assume that $\tau \in Aut_n(\mathbb{Z}G)$ induces τ_K on $\mathbb{Z}G/K$.

Suppose that τ_K fixes the class sums of all elements of prime power order. Then for each p, which does not divide $|K|$, τ fixes the class sums of p-elements.

PROOF: The group epimorphism π induces a ring homomorphism from $\mathbb{Z}G$ onto $\mathbb{Z}G/K$, also denoted by π. Clearly π maps class sums into multiples of class sums. By Lemma 1.3 it follows that π yields a bijection between the class sums of p-elements. Consequently τ must fix these class sums, if τ_K has this property. $\qquad\square$

1.5. Theorem. *Let M, N be normal subgroups of G with the following properties.*

a) $I(M) \cdot \mathbb{Z}G$ and $I(N) \cdot \mathbb{Z}G$ are characteristic ideals of $\mathbb{Z}G$.

b) $M \cap N = 1$ and $(|M|, |N|) = 1$.

c) Each normalized automorphism τ_M, τ_N rsp. of $\mathbb{Z}G/M$, $\mathbb{Z}G/N$ rsp. may be written with respect to each group basis as

$$\alpha_M \cdot \sigma_M, \alpha_N \cdot \sigma_N \text{ rsp.}$$

where α_M, α_N rsp. is induced from a group automorphism of the group basis and σ_M, σ_N rsp. is a normalized ring automorphism fixing the class sums of conjugacy classes of each element of prime power order.

d) $G/M \cdot N$ is abelian.

Then each normalized ring automorphism of G is with respect to each group basis the product of an automorphism induced from a group automorphism of the group basis and a normalized ring automorphism, which fixes the class sums of conjugacy classes of each element of prime power order.

PROOF: Let H be a group basis of $\mathbb{Z}G$. We show that the assumptions a), b) and d) are also satisfied with respect to H. By Theorem V.1 (e) we have a normal subgroup correspondence between H and G induced from the class sum correspondence. Thus M, N correspond to normal subgroups M_H and N_H of H. Moreover $I(M)\cdot\mathbb{Z}G = I(M_H)\cdot\mathbb{Z}G$ and $I(N)\cdot\mathbb{Z}G = I(N_H)\cdot\mathbb{Z}G$, since the class sums of M, M_H and N, N_H rsp. coincide. Consequently H satisfies assumption a) with M_H and N_H. Since the normal subgroup correspondence preserves the orders of the normal subgroups and establishes a lattice isomorphism, it follows that assumption b) carries over to H. Finally $H/M_H \cdot N_H$ is abelian, since $M_H \cdot N_H$ is precisely the kernel of the ring homomorphism from $\mathbb{Z}G$ onto $\mathbb{Z}G/M \cdot N$ restricted to H.

The calculation

$$m \cdot n - 1 = (m-1) \cdot (n-1) + m - 1 + n - 1$$

shows that $I(M \cdot N) \cdot \mathbb{Z}G$ is a characteristic ideal of $\mathbb{Z}G$, because $I(M) \cdot \mathbb{Z}G$ and $I(N) \cdot \mathbb{Z}G$ have this property.

Now we are in the position to modify (accordingly to our strategy) a given $\tau \in Aut_n(\mathbb{Z}G)$ with respect to each group basis B of $\mathbb{Z}G$ by a group automorphism of B such that the modification induces on $\mathbb{Z}G/M$ and on $\mathbb{Z}G/N$ ring automorphisms fixing the class sums. Therefore we may apply Corollary 1.4. Note that the assumption $(|M|, |N|) = 1$ implies that we can for each prime p dividing $|G|$ apply Corollary 1.4 either with respect to M or with respect to N. Thus we get the conclusion of the theorem for all elements of prime power order. □

1.6. Theorem.

Assume that $[G,G]$ is nilpotent, then G satisfies ZC-Variation 2.

PROOF: If the Fitting subgroup[32] $F(G)$ is a p-group, then by VI.2 of section VI even the Zassenhaus-conjecture holds for G. If G is abelian, then the class sum correspondence shows that again the Zassenhaus-conjecture holds for G.

If $F(G)$ is not a p-group and G is not abelian, then put $M = O_p(G)$ and $N = O_{p'}(G)$ where $p \in \pi([G,G])$. The normal subgroup correspondence shows that $I(M) \cdot \mathbb{Z}G$ and $I(N) \cdot \mathbb{Z}G$ are characteristic ideals of $\mathbb{Z}G$. The nilpotence of $[G,G]$ and the choice of p yields that $G/M \cdot N$ is abelian. Clearly $M \cap N = 1$ and $(|M|,|N|) = 1$. By induction on the group order we may assume that hypothesis c) of Theorem 1.5 is satisfied. Hence Theorem 1.5 yields that Aut-Variation 2 holds for G.

Finally, since the class of groups with nilpotent commutator subgroup is closed under direct products, we obtain by Theorem 2.4 of section X that ZC-Variation 2 holds for groups with nilpotent commutator subgroup. □

Theorem 1.6 yields immediately Theorem 1.1. For, if a group has a nilpotent commutator subgroup then it belongs to the class NA. Conversely, if a group has a normal nilpotent subgroup such that the quotient is abelian, then its commutator subgroup has to be nilpotent, since subgroups of nilpotent groups are nilpotent.

§2   Sylow subgroups

One of the well known problems posed by R.Brauer was that of recognizing from the character table of a finite group whether or not, for a given prime p, the Sylow p-subgroups of the group were abelian [Bra; 63, Problem 12]. W.Feit stimulated the work on this problem by stating it as something which should be answerable with the classification of the finite simple groups [Fe; 80].

The special case when $p = 2$, has been solved by A.Camina and M.Herzog [C-H; 80]. Their result depends only on the classification of simple groups with abelian Sylow 2-subgroups. The positive answer to the general case has its origin in the chief series result for integral group rings [K-L-S-T; 90], cf. V.1 (f) of section V. Following Feit's suggestion, the proof uses the classification of the finite simple groups and is given in [Ki; 91], and in a more general way in [K-S; 91]. In fact it turns out that the answer is even positive in the case of abelian Hall subgroups[33].

---

32   The Fitting subgroup is the largest normal nilpotent subgroup of a finite group.
33   A Hall π-subgroup of a finite group G is a π-subgroup which contains for each $p \in \pi$ a Sylow p-subgroup of G.

2.1. Theorem. *[Ki; 91] [K-S; 91]*

The character table of a finite group G determines, whether or not, for a given set π of primes , G has abelian Hall π-subgroups. If the Hall π-subgroups are abelian, then they are determined up to isomorphism.

PROOF (SKETCH):

The first step is to establish the result for abelian Sylow subgroups.

By transfer it follows for a group G with abelian Sylow subgroups that $S \cap [G, G] \cap Z(G) = 1$.

Assume that the generalized Fitting group $F^*(G)^{34}$ is a simple group S. Then, by [G-L; 83, 7.10], if p is a prime dividing $|S|$ and $|G/S|$ Sylow p-subgroups of G are non-abelian.

This leads to the result that a finite group G with abelian Sylow p-subgroups has the following structure.

G has a normal series

$$1 < N < M < G$$

with $N = O_{p'}(G)$, M is a direct product of simple groups L_i and $O_p(M/N)$, G/M is a p'-group.

Note that this structure may be detected by the normal subgroup lattice and the orders of the normal subgroups. This is available when the character table is given by 2.1 of section V. Moreover, if G has such a structure, we get from Theorem 2.2 of section V that the simple groups L_i are determined by the character table of G up to isomorphism. Consequently it may be decided whether they have abelian Sylow p-subgroups or not.

Reducing mod $O_{p'}(G)$ and then mod the L'_i s it remains to show for a group with normal Sylow subgroup P that the character table determines whether P is abelian. Now in this situation P is abelian if, and only if, the order of the centralizer of each p-element is divisible by $|P|$. The latter may be easily checked with the orthogonality relations. Finally the isomorphism type of P is determined inductively reducing modulo the socle of P.

In order to extend the result to Hall subgroups we use the following group-theoretical fact.

2.2. Proposition. *[Ki; 91, 4.10(i)] Let G be a finite group and let π be a set of primes. Then the following are equivalent.*

a) For each $p \in \pi$ and each $q \in \pi \backslash \{p\}$ the centralizer $C_G(x)$ contains for each p-element x a Sylow q-subgroup of G and each chief factor of G has nilpotent Hall π-subgroups.

b) G has nilpotent Hall π-subgroups.

---

34   The generalized Fitting subgroup is the largest quasinilpotent normal subgroup of a finite group. It consists of all group elements which act on each chief factor as inner automorphism.

By 2.1 and 2.2 part a) of the above proposition is determined by the character table of a group. So part b) follows and together with the previous this completes the proof of Theorem 2.1. $\square$

The analogous questions in the context of the isomorphism problem of integral group rings have been studied extensively during the last years and are still one main topic in the actual research with respect to the isomorphism problem and related questions with respect to block theory, cf. e.g. [Sc; 87]. Note in this context Scott's result, Theorem VI.3 of section VI. As the character table of a finite group G is determined by its integral group ring $\mathbb{Z}G$, the determination of abelian Sylow and Hall subgroups up to isomorphism is a consequence of Theorem 2.1. In the context of the isomorphism problem at least a little bit more is valid.

2.3. Theorem. *[K-S; 91] The integral group ring $\mathbb{Z}G$ of a finite group G determines hamiltonian[35] Hall subgroups up to isomorphism.*

Clearly Theorem 2.3 does not hold with respect to character tables. Note that Theorem 2.3 makes no restriction on G. Much more, however, holds in the context of the isomorphism problem, if G is not arbitrary. The key result for this is Theorem VI.2 of section VI. This Theorem has many consequences with respect to the determination of Sylow and Hall subgroups. One is, that normal Sylow subgroups are determined. Another one is, noted in [Sc; 87], that Sylow subgroups of a soluble group are determined, even more in [K-R; 91] it has been proved that a Sylow-like theorem for the group of normalized units $V(\mathbb{Z}G)$ holds, provided G is soluble.

2.4. Theorem. *Let G be a finite soluble group and H be a group basis of $\mathbb{Z}G$. Then Sylow p-subgroups of G and H are conjugate by a unit of $\mathbb{Q}G$.*

PROOF (SKETCH): Consider the exact sequence

$$0 \to I(O_{p'}(G)) \cdot \mathbb{Z}G \to \mathbb{Z}G \xrightarrow{\pi} \mathbb{Z}G/O_{p'}(G) \to 0$$

Since $O_{p'}(G/O_{p'}(G)) = 1$ and G is soluble, we may apply Theorem VI.2 of section VI and obtain that the Zassenhaus Conjecture is valid for $\mathbb{Z}G/O_{p'}(G)$. Thus we get an isomorphism between $G/O_{p'}(G)$ and $\pi(H)$ fixing the class sums. By Lemma 1.3 we can lift this isomorphism to a bijective map σ from G to H which fixes the class sums and which is restricted to a Sylow p-subgroup P of G a homomorphism.

Let B be a Wedderburn component of $\mathbb{C}G$ and let π_B be the projection from $\mathbb{Z}G$ into B. Now the representations π_B and $\sigma\pi_B$ of P afford the same $\mathbb{C}$-characters, since σ fixes the class sums. Hence $P\pi_B$ and $P\sigma\pi_B$ are conjugate in B. Obviously puzzling the conjugation componentwise together we get that P and $P\sigma$ are conjugate in $\mathbb{C}G$.

---

35   A group is called hamiltonian, if each subgroup is normal.

Finally a standard Noether-Deuring argument shows that this conjugation takes already place in $\mathbb{Q}G$.

□

2.5. Theorem. *[K-S; 91] Let G be a soluble group. Then $\mathbb{Z}G$ determines nilpotent Hall subgroups of G up to isomorphism.*

PROOF: The reduction mod $O_{p'}$ used at the beginning of the proof of Theorem 2.4 shows that group bases of $\mathbb{Z}G$ have isomorphic Sylow subgroups. Now Proposition 2.2 gives immediately the result.   □

2.6. Remarks. The sketch of the proofs of Theorem 2.1 makes also transparent that essentially group-theoretical methods are used.

In [K-S; 91] the results are posed in the more general context of a Jordan-Hölder- , a Jordan-Hölder-Sylow and a Jordan-Hölder-Brauer class correspondence rsp. The rough idea of a Jordan-Hölder class correspondence is to link groups G and H only by a bijection which maps conjugacy classes into conjugacy classes and which behaves well for factor groups.

Class correspondences of Jordan-Hölder-Sylow type additionally preserve for each given prime the set of p-elements, those of type Jordan-Hölder-Brauer type are compatible with the power map on the conjugacy classes.

Note that, if G and H have the same character table, then they are in Jordan-Hölder-Sylow correspondence. If $\mathbb{Z}G$ and $\mathbb{Z}H$ are isomorphic, then G and H are in class correspondence of type JHB. For a detailed discussion see [K-S; 91].

Some special types of non-abelian Sylow subgroups are easily characterized by the integral group ring provided the abelian ones are determined [K-S; 91]. This is in particular the case for Sylow subgroups which have a cyclic normal subgroup of index p. It should be mentioned that Sylow 2-subgroups of this type may already be recognized by the use of modular group algebras (using the Auslander-Reiten quiver) [Er; 90]. Also we mention that Bessenrodt [Bes; 89] has shown that the isomorphism type of a Sylow p-subgroup known to be abelian can already be determined by use of a group ring with a field of characteristic p. Her argument then clearly does not invoke the classification of the finite simple groups.

Finally in order to indicate another important relationship to modular representations we remark that Michler has pointed out that a positive answer to Brauer's question may be derived from Brauer's well-known height-zero conjecture. However this conjecture is still an open question [Mic; 86], although a good piece of progress has been made within the last years to a proof of this conjecture mainly within the "DFG-Schwerpunktsprogramm Darstellungstheorie endlicher Gruppen und endlich-dimensionaler Algebren".

XIII Clifford theory revisited

The arguments here are quite condensed and use a fair amount of integral representation theory. Since it is essential only to construct counter examples, it can be skipped by the unfamiliar reader.

We shall present here an integral form of Clifford's theory, which is crucial for the counterexample to the Zassenhaus conjecture in Section IX and also in the example of locally isomorphic group rings. In case the base ring has a quotient field which is a splitting field for G and all of its subgroups, the argument is due to [Sch; 88 1, Sch; 83, Sch; 88 2][36]. However, the version we prove here holds generally.

XIII.1. Assumption. Let R be a complete Dedekind domain of characteristic zero with maximal ideal $\wp$ and $p\mathbb{Z} \in Spec(\mathbb{Z})$ such that $p \cdot R \neq R$ and assume that $R/\wp$ is finite. By K we denote the quotient field of R. G is a finite group and N is a normal subgroup of G with $|N| \cdot R = R$. It should be *noted, that we do not require – as is usually done – that K is a splitting field for G and all of its subgroups.*

We shall prove the following

XIII.2. Theorem. *Assume that*

- *N is a normal subgroup of the finite group G,*
- *with $|N| \cdot R = R$ and $(|N|, |G : N|) = 1$.*
- *Let M be an irreducible RN–lattice with $End_{RN}(M) = S$ a finite unramified extension of R;*
- *let $n = dim_S(M)$.*
- *Let $I(M)$ and $I_S(M)$ be the inertia groups of M as RN–module and SN–module resp.*
  Then $I_S(M) \trianglelefteq I(M)$ with quotient $T(M)$ which acts as group of R–Galois automorphisms on S.
- *Assume furthermore that $I_S(M)$ has a complement $T_0(M)$ in $I(M)$.*
- *Denote by $SI_S(M)/N^\circ$ the $RI(M)$–module, where $I_S(M)$ acts by left multiplication and $T_0(M)$ acts by conjugation.*
- *S° is the $RT_0(M)$-module, where $T_0(M)$ acts via $T(M)$ as Galois automorphism.*

Then the group ring RG contains a ring direct summand of the form

$$B := Mat_{|G:I_S(M)| \cdot n}(H^0(T_0(M), RI_S(M)/N^\circ) \otimes_R S^\circ)$$

where $H^0(-, -)$ is the fixed points functor.

The proof will proceed in several steps.

XIII.3. Proposition.

$$RN \simeq \prod_{1 \leq i \leq s} Mat_{n_i}(R_i),$$

where $\{R_i\}_{1 \leq i \leq s}$ *are unramified extensions of* R *of finite degree. In particular, if* $R = S$ *; i.e.* K *is a splitting field for* N, *then* $R_i = R$ *for all* $1 \leq i \leq s$.

PROOF OF PROPOSITION XIII.3: Let $I\!F = R/\wp \cdot R$ be the residue field of R. Then $I\!F N$ is semi simple by Maschke's theorem, and so $rad(RN) = \wp \cdot RN$. On the other hand every RN–lattice is projective – apply again Maschke's argument, and two RN–lattices are isomorphic if and only if they are isomorphic over K^{37}. Thus RN is a hereditary order. However since $\wp \cdot RN = rad(RN)$, the group ring must have the above form ([Re; 75, (17.3) Theorem] (no skew-fields arise, because of [Re; 75, (14.3) Theorem]).
q.e.d. Proposition XIII.3   □

(51) We now *fix for the rest of this section* an indecomposable RN–lattice M which corresponds to a block B of RN, $B \simeq Mat_n(S)$ for $S = R_i$ and $n = n_i$ as stated above for some i ; i.e. $End_{RN}(M) = S$, where by XIII.3 S with field of fractions L is a finite unramified extension of R.

M can also be viewed as SN–module; which we shall denote by M_S. If we view M as SN–module, we get the ordinary Clifford theory; however, when we view M as RN–module, then the situation is quite different.

Note that also

$$End_{SN}(M_S) = S.$$

XIII.4. Definition. Let M be as in (51). Since N is normal, we can form to $g \in G$ the *conjugate module* for $g \in G$, ${}^g M$, which is M with the action $n \cdot_{{}^g M}$ $m = {}^g n \cdot m$ [38]. We note that ${}^g M \simeq_{RN} M$ and ${}^g M_S \simeq_{SN} M_S$ if $g \in N$ [39].

In order to *define the inertia group* of M, we have to be very careful, whether we look at M as RN– or as SN–module.

The *inertia group* of M as RN–module is defined as

$$I(M) = \{g \in G : \ {}^g M \simeq_{RN} M\}.$$

The *inertia group* of M_S as SN–module is defined as

$$I_S(M) = \{g \in G : \ {}^g M_S \simeq_{SN} M_S\}^{40}.$$

We have $I(M) \geq I_S(M_S)$, since $R \subseteq S$.

---

37   Use the decomposition map[C-R; 62]
38    ${}^g n = g \cdot n \cdot g^{-1}$
39   send $n \cdot m$ to m
40    $I(M)$ and $I_S(M)$ are surely subgroups of G

XIII.5. Definition. *Let* $G = \{g_i\}_{1 \leq i \leq s}$. *For each* $g_i \in I_S(M_S)$ *we fix an SN-isomorphism*

$$\phi(g_i) \ : \ {}^{g_i}M_S \longrightarrow M_S.$$

Then $\phi(g_i) \in End_S(M)$, *and we have*

$$\phi(g_i)({}^{g_i}n \cdot m) = n \cdot \phi(g_i)(m) \ \text{for all } n \in N \ \text{ and all } m \in M,$$

i.e.

$$\phi(g_i) \ {}^{g_i}n \ \phi(g_i)^{-1} \ = \ n \ \text{for each } n \in I\!N,$$

viewing n *in its action on* M_S; *i.e. interpreting* $n \in N$ *as an element in* $End_S(M)$. *Since* M_S *is irreducible,* $\phi(g_i)$ *is uniquely determined up to scalar multiples in* S.

For each $g_i \in I(M)$ *we fix an RN-isomorphism*

$$\psi(g_i) \ : \ {}^{g_i}M \longrightarrow M.$$

Then $\psi(g_i) \in End_R(M)$, *and we have*

$$\psi(g_i)({}^{g_i}n \cdot m) = n \cdot \psi(g_i)(m) \ \text{for all } n \in N \ m \in M,$$

i.e.

$$\psi(g_i) \ {}^{g_i}n \ \psi(g_i)^{-1} \ = \ n \ \text{for each } n \in N,$$

viewing n *in its action on* M; *i.e. interpreting* $n \in N$ *as an element in* $End_S(M)$. *Since* M *is irreducible,* $\psi(g_i)$ *is uniquely determined up to scalar multiples in* S.

We choose these homomorphisms such that

$$(52) \qquad\qquad \text{for } g_i \in I_S(M) \text{ we have } \phi(g_i) = \psi(g_i).$$

It also should be noted that for $g_i \in I(M) \setminus I_S(M)$ the homomorphism $\psi(g_i)$ is definitely not an SN–isomorphism.

We refer now to section XIV 3.1 for an interesting example concerning the change of the ring structure, if one extends the coefficient domain.

XIII.6. Note. Assume for the moment that $G = I(M)$ and put

$$\mu(g_i, g_j) = \psi(g_i)^{-1} \cdot \psi(g_j)^{-1} \cdot \psi(g_i \cdot g_j).$$

Note that this measures how far ψ is from being a homomorphism. [41] Let $g_j \cdot g_i = g_k$. Then for every $n \in N$ we have

$$
\begin{aligned}
\mu(g_j, g_i) \cdot {}^{g_k}n \cdot \mu(g_j, g_i)^{-1} &= \psi(g_j)^{-1} \cdot \psi(g_i)^{-1} \cdot \psi(g_k) \cdot {}^{g_k}n \cdot \psi(g_k)^{-1} \cdot \psi(g_i) \cdot \psi(g_j) \\
&= \psi(g_j)^{-1} \cdot \psi(g_i)^{-1} \cdot n \cdot \psi(g_i) \cdot \psi(g_j) \\
&= \psi(g_j)^{-1} \cdot {}^{g_i}n \cdot \psi(g_j) = {}^{g_j \cdot g_i}n = {}^{g_k}n.
\end{aligned}
$$

---

41   Since we have written maps on the left, ψ is a homomorphism iff $\psi(gh) = \psi(h)\psi(g)$.

Thus $\mu(g_i, g_j)$ centralizes N in its action on the irreducible RN–module M, and hence is scalar multiplication with a diagonal matrix with entries in S^*, the group of units in S – recall that $S = End_{RN}(M)$.

XIII.7. Proposition. *Assume that $I(M) = G$. As above we put*

$$\mu(g_j, g_i) := \psi(g_j)^{-1} \cdot \psi(g_i)^{-1} \cdot \psi(g_j \cdot g_i).$$

Then

$$\mu : \ G \times G \longrightarrow S^*$$

is a 2-cocycle of G with coefficients in the G–module S^*, which is uniquely determined up to 2-coboundaries. It vanishes on N and hence defines a unique element in $H^2(G/N, S^*)$.

Moreover, if $(|I_S(M)/N|, |N|) = 1$, then μ restricted to $I_S(M)$ is a 2-coboundary, and consequently, M can be extended to an $SI_S(M)$-module. In particular, we may assume, that μ is the identity on $I_S(M)$.

Before we come to the proof, *we repeat* – for the readers convenience – the definition of a *2-cocycle*, a *2-coboundary* and the *second cohomology* group.

Most naturally a 2-cocycle arises as obstruction of a set theoretical splitting of a short exact sequence of groups with abelian kernel A (written multiplicatively) and quotient G to be a homomorphism. This kernel is given a G–module structure by conjugating with some inverse image. This operation is well defined since A is abelian. If f is a set theoretical splitting, then

$$f(g) \cdot f(h) = f(gh) \cdot \sigma(g, h).^{42}$$

for a unique $\sigma(g, h) \in A$. This follows from the fact that G is a group. Let us test the associativity:

$$
\begin{aligned}
f(g)(f(h)f(k)) &= f(g) \cdot (f(hk) \cdot \sigma(h, k)) \\
&= f(ghk) \cdot \sigma(g, hk) \cdot \sigma(h, k) \\
(f(g)f(h))f(k) &= f(gh) \cdot \sigma(g, h) \cdot f(k) \\
&= f(gh) \cdot f(k) \cdot (\, {}^{f(k)^{-1}}\sigma(g, h)) \\
&= f(ghk) \cdot \sigma(gh, k) \cdot (\, {}^{f(k)^{-1}}\sigma(g, h)).
\end{aligned}
$$

Multiplying by $f(ghk)^{-1}$ from the left side one obtains – viewing A as G–module

$$\sigma(g, hk) \cdot \sigma(h, k) = \sigma(gh, k) \cdot \, {}^{f(k)^{-1}}\sigma(g, h)$$

The function $\sigma : G \times G \longrightarrow A$ is then called a 2-cocycle of G with values in the G–module A.

---

42   This is the usual definition contrary to our use of a 2–cocycle above which arises, since we have written maps on the left.

What happens, if one takes another splitting f? This discussion leads to the definition of a 2–coboundary.

For simplicity we assume the extension of groups to be split. Then f can be chosen as group homomorphism. If one chooses another set theoretic splitting f' with associated 2–cocycle σ, then we calculate:

$$
\begin{aligned}
f(g)^{-1} \cdot f'(g) &=: \mu(g) \\
f'(gh) \cdot \sigma(g,h) &= f'(g)f'(h) \\
\sigma(g,h) &= f'(gh)^{-1}f'(g)f'(h) \\
&= \mu(gh)^{-1} \cdot f(gh)^{-1} \cdot f(g) \cdot \mu(g) \cdot f(h) \cdot \mu(h) \\
&= \mu(gh)^{-1} \cdot f(gh)^{-1} \cdot f(g) \cdot f(h) \cdot {}^{f(h)^{-1}}\mu(g) \cdot \mu(h) \\
&= \mu(gh)^{-1} \cdot {}^{f(h)^{-1}}\mu(g) \cdot \mu(h),
\end{aligned}
$$

since f is a group homomorphism. A function $\sigma : G \times G \longrightarrow A$ is called a *2–coboundary*, if there is a function $\mu : G \longrightarrow A$ such that

$$
\sigma(g,h) = \mu(gh)^{-1} \cdot {}^{h^{-1}}\mu(g) \cdot \mu(h)
$$

for all $g, h \in G$.

Note that $g \in G$ acts on $a \in A$ by conjugation with $f(g)$. We have written this action as ${}^{f(g)}a = {}^{g}a$.

Direct computation shows, that every 2–coboundary is a 2–cocycle and both sets form abelian groups (this uses heavily that A is abelian).

The *second cohomology group* $H^2(G, A)$ of a group G with values in a G–module A is defined to be the group of 2–cocycles modulo the group of 2–coboundaries. This group classifies the group extensions in the described way.

XIII.8. Lemma. *Assume for the moment, that $I(M) = G$. With the notation of XIII.4 we have that $I_S(M)$ is a normal subgroup of $I(M)$, and the quotient $T(M) = I(M)/I_S(M)$ injects naturally into $Gal(S/R)$, the group of R-automorphisms of S. Note, that S is a Galois extension of R with cyclic Galois group, S being unramified over R [Has; 49][43]. Thus $T(M)$ is a cyclic group. By R_0 we denote the subring of S fixed by T. Via the action of T on S, S and also the multiplicative group S^* become $I(M)$-modules. The action is given – as the proof will show – by*

$$
{}^{g}s = \psi(g) \cdot s \cdot \psi(g)^{-1}.
$$

PROOF OF LEMMA XIII.8: The RN–isomorphism $\psi(g) : {}^{g}M \longrightarrow M$ induces an isomorphism $\rho(g) : End_{RN}({}^{g}M) \longrightarrow End_{RN}(M)$ via

$$
\alpha \in End_{RN}({}^{g}M) \longrightarrow \psi(g) \cdot \alpha \cdot \psi(g)^{-1} \in End_{RN}(M).
$$

---

43   The Galois group is isomorphic to that of the field extensions corresponding to the residue fields of R and S and is therefore generated by the Frobenius homomorphism.

Since $End_{RN}(^gM) = S = End_{RN}(M)$, the map $\rho(g)$ can be interpreted as an automorphism – also denoted by $\rho(g) : S \longrightarrow S$.

In order to complete the proof we show:

XIII.9. Claim. *The map*

$$\rho : I(M) \longrightarrow Aut(S),$$

$$\rho(g) : s \longrightarrow \psi(g) \cdot s \cdot \psi(g)^{-1}$$

is a homomorphism of groups with kernel $I_S(M)$, and it has image in $Gal(S/R)$.

PROOF OF CLAIM XIII.9: Since $\psi(g)$ is an RN–isomorphism, the automorphism $\rho(g)$ is surely R–linear. Moreover, if $g \in I_S(M)$, then $\psi(g)$ is S–linear, and then $\rho(g) = id_S$ is the identity. Conversely, if $\rho(g) = id_S$ we have that $\psi(g)$ commutes with all of S and thus is S–linear.

It remains to show, that ρ is a homomorphism of groups. However, for $\alpha \in End_{RN}(^{gh}M)$ we have

$$
\begin{aligned}
\rho(gh)(\alpha) &= \psi(gh) \cdot \alpha \cdot \psi(gh)^{-1} \\
&= \psi(h) \cdot \psi(g) \cdot \mu(g,h) \cdot \alpha \cdot \mu(g,h)^{-1} \cdot \psi(g)^{-1} \cdot \psi(h)^{-1} \text{ by XIII.6.}
\end{aligned}
$$

But, α is scalar multiplication with an element $s_0 \in S$, and since $\mu(-,-)$ also lies in the commutative ring S, we conclude $\rho(gh) = \rho(h) \cdot \rho(g)$ and ρ is a group homomorphism. q.e.d.Claim XIII.9   □

We now return to the proof of Lemma XIII.8: Since $I_S(M)$ is the kernel of the homomorphism ρ, $I_S(M)$ is normal in $I(M)$ and the quotient $T = I(M)/I_S(M)$ is cyclic, since $Gal(S/R)$ is cyclic. q.e.d. Lemma XIII.8   □

PROOF OF PROPOSITION XIII.7: By XIII.8, G acts on S^* by conjugation with $\psi(g)$, and thus we have to show that μ from XIII.6 is a multiplicative 2–cocycle:

Recall:
$$\mu(g,h) = \psi(g)^{-1} \cdot \psi(h)^{-1} \cdot \psi(gh) \in S^*$$

and

$$^{\rho(g)}s = \psi(g) \cdot s \cdot \psi(g)^{-1}, s \in S.$$

We have to show:

$$^{g^{-1}}\mu(h,k) \cdot \mu(g,hk) = \mu(g,h) \cdot \mu(gh,k)^{44}$$

However,

$$\psi(g \cdot (hk)) = \psi(hk) \cdot \psi(g) \cdot \mu(g,hk)$$

---

44   Note that our multiplication is contravariant

$$
\begin{aligned}
&= \psi(k) \cdot \psi(h) \cdot \mu(h, k) \cdot \psi(g) \cdot \mu(g, hk) \\
\psi((gh) \cdot k) &= \psi(k) \cdot \psi(gh) \cdot \mu(gh, k) \\
&= \psi(k) \cdot \psi(h) \cdot \psi(g) \cdot \mu(g, h) \cdot \mu(gh, k).
\end{aligned}
$$

Since $\psi(g \cdot (hk)) = \psi((gh) \cdot k)$, we get

$$
\mu(h, k) \cdot \psi(g) \cdot \mu(g, hk) = \psi(g) \cdot \mu(g, h) \cdot \mu(gh, k),
$$

since we can cancel.

The uniqueness is shown as follows: Take another RN–isomorphism $\chi(g): {}^{g}M \longrightarrow M$. Then $\chi(g) \doteq \psi(g) \cdot s_g$ for an automorphism

$$
s_g \in End_{RN}(M) = S,
$$

and one gets an associated 2–cocycle

$$
\begin{aligned}
\nu(h, g) &= \chi(h)^{-1} \cdot \chi(g)^{-1} \cdot \chi(gh) \\
&= s_h^{-1} \cdot \psi(h)^{-1} \cdot s_g^{-1} \cdot \psi(g)^{-1} \cdot \psi(gh) \cdot s_{gh} \\
&= s_h^{-1} \cdot {}^{\psi(h)^{-1}}(s_g^{-1}) \cdot \psi(h)^{-1} \cdot \psi(g)^{-1} \cdot \psi(gh) \cdot s_{gh} \\
&= s_h^{-1} \cdot {}^{\psi(h)^{-1}}(s_g^{-1}) \cdot \mu(h, g) \cdot s_{gh} \ ;
\end{aligned}
$$

however, $\mu(g, h) \in S$ commutes with s_{gh}, and so $\nu(h, g) = s_g^{-1} \cdot {}^{\psi(g)^{-1}}(s_h^{-1}) \cdot s_{gh} \cdot \mu(h, g)$. Hence μ and ν differ by a 2–coboundary, and so μ gives rise to a unique element in $H^2(G, S^*)$.

XIII.10. Claim. μ *gives rise to a unique element* $\overline{\mu}$ *in* $H^2(G/N, S^*)$ *defined by*

$$
\overline{\mu}(Ng, Nh) =: \mu(g, h).
$$

PROOF: We have to show that this is well defined. Let

$$
G = \bigcup Ng_i
$$

be the decomposition into cosets. First we can arrange the isomorphisms ψ such that $\psi(ng_i) = \psi(g_i) \cdot \psi(n)$ for all $n \in N$ and all i. In fact we had that

$$
\psi(n)(m) = n^{-1}m
$$

for all $m \in M, n \in N$ (cf. XIII.4), but then $\psi(n)$ also induces an isomorphism from ${}^{ng_i}M$ to ${}^{g_i}M$. and thus $\psi(g_i) \cdot \psi(n)$ is an RN–isomorphism from ${}^{ng_i}M$ to M.

By definition we have the relation

$$
\psi(g) \cdot {}^{g}n = n \cdot \psi(g)
$$

for all $g \in G, n \in N$. Let us now compute for $n_1, n_2 \in N$:

$$
\begin{aligned}
\mu(n_1 g_i, n_2 g_j) &= \psi(n_1 g_i)^{-1} \psi(n_2 g_j)^{-1} \psi(n_1 g_i n_2 g_j) \\
&= \psi(n_1)^{-1} \psi(g_i)^{-1} \psi(n_2)^{-1} \psi(g_j)^{-1} \psi(n_1 {}^{g_i} n_2 g_i g_j) \\
&= \psi(n_1)^{-1} \psi(g_i)^{-1} \psi(n_2)^{-1} \psi(g_j)^{-1} \psi(g_i g_j) \psi(n_1 {}^{g_i} n_2) \\
&= \psi(n_1)^{-1} ({}^{g_i}\psi(n_2))^{-1} \mu(g_i, g_j) \psi({}^{g_i} n_2) \psi(n_1) \\
&= \mu(g_i, g_j),
\end{aligned}
$$

since $\psi({}^{g_i} n_2) = {}^{g_i}\psi(n_2)$ is multiplication by ${}^{g_i} n_2^{-1}$ and since μ has values in S.

This proves the claim. q.e.d.   □

XIII.11. Note. If $\overline{\mu}$ is a coboundary then so is μ as one sees from the definition.

In order to complete the proof, we recall from XIII.5, that for $g \in I_S(M)$, $\psi(g) = \phi(g)$ (cf. (52)) is an SN–isomorphism; in particular, $\phi(g)$ centralizes S. We put $E = I_S(M)/N$ and assume now, that $(|N|, |E|) = 1$. We then have for $x, y \in E$ – note that μ vanishes on N –

$$
\mu(x, y) = \psi(x)^{-1} \cdot \psi(y)^{-1} \cdot \psi(x \cdot y).
$$

We interpret this as an equation of S–linear maps on M. Taking determinants we get with $d = dim_L(LM)$ and L being the quotient field of S:

$$
\mu(x, y)^d = det(\psi(x)) \cdot det(\psi(y)) \cdot det(\psi(x \cdot y))^{-1}.
$$

Now all factors lie in S^*, and the right hand side is exactly the condition for μ^d being a 2–coboundary as one sees from the definition above. (Note that $\phi(g)$ centralizes S.)

On the other hand, the cohomology group $H^2(E, S^*)$ is annihilated by $|E|$[45]. Since the degrees of the characters divide the group order, the dimension d divides $|N|$. However, $|E|$ and $|N|$ are relatively prime, and so μ must be a coboundary, since it is annihilated by $|N|$ and $|E|$. Say for $g, h \in I_S(M)$, $\mu(g, h) = s_g \cdot s_h \cdot s_{gh}^{-1}$ – note that $\phi(g)$ centralizes S. If we now replace $\psi(g)$ by $\chi(g) = \psi(g) \cdot s_g$, then $\chi(g)$ is an SM–isomorphism from ${}^g M$ to M, and an easy calculation – as above – shows that $\chi : I_S(M) \longrightarrow End_S(M)$ is a homomorphism; i.e. M extends to an $SI_S(M)$–module. The modified cocycle associated to χ vanishes on $I_S(M)$.

q.e.d. Proposition XIII.7   □

XIII.12. Claim. *Let $\{g_i\}_{1 \leq i \leq t}$ be a set of coset representatives of $I_S(M)$ in $I(M)$. Then the RG–homomorphisms $\{\psi(g_i)\}_{1 \leq i \leq t}$ are in fact R_0–isomorphisms, where R_0 is the fixed ring of S under $T(M)$ acting via ρ. Moreover, the SN–module ${}^{g_i} M$ is the Galois conjugate module to M under $\rho(g_i)$. In addition, as $SI(M)$–module we have an isomorphism $S \otimes_{R_0} M \simeq_{SI(M)} M \uparrow_{I_S(M)}^{I(M)}$.*

---

45   use a "Maschke type argument"

PROOF OF CLAIM XIII.12: By the definition of $\psi(g_i): \ {}^{g_i}M \longrightarrow M$ we have for $s \in S$

$$s \cdot \psi(g_i) = \psi(g_i) \cdot {}^{\psi(g_i)^{-1}}s \ ;$$

however, ${}^{\psi(g_i)^{-1}}s = {}^{g_i^{-1}}s$ is the Galois action. In particular, if $s \in R_0$, then this shows, that $\psi(g_i)$ is an R_0N–isomorphism.

Consequently, we do not loose anything, if we assume for the time being, that $R = R_0$. Recall, that M is an RN–module corresponding to the simple component $B = (S)_n$.

Then $S \otimes_R M$ decomposes into the Galois conjugate modules ${}^{\rho(g_i)}M$ corresponding to the t simple components of

$$S \otimes_R B \simeq \prod_1^t (S)_n, \text{ with } t = |T(M)|.$$

These Galois conjugate modules are all RN–isomorphic to M; but as SN–modules they are non isomorphic. Now, the conjugate SN–modules $\{{}^{g_i}M\}_{1 \leq i \leq t}$ have the same property. But the group ring SN has exactly t non isomorphic modules, which become isomorphic as RN–modules. Our construction of ρ then shows, that ${}^{\rho(g_i)}M \simeq_{SN} {}^{g_i}M$, as claimed. This also shows, that

$$S \otimes_R M \simeq_{SI(M)} M \uparrow^{I(M)}_{I_S(M)} .$$

q.e.d.Claim XIII.12   □

XIII.13. Lemma. *We keep the assumptions of XIII.8, and assume that M is an $I_S(M)$–module. Assume furthermore that $G = I(M)$. Then M can be extended to an RG–module, also denoted by M, and the induced module $M \uparrow^G_{I_S(M)}$ decomposes as RG–module into t copies of M where $t = |I(M) : I_S(M)|$.*

PROOF OF LEMMA XIII.13: Let R_0 be the fixed ring of S under the Galois action of T. We have seen in XIII.12, that – up to now – there is no loss of generality, if we *assume* that $R = R_0$.

For the sake of simplicity we put $I := I_S(M)$ and $M_I^G := M \uparrow_I^G$. Because of XIII.12 and its proof the SG–lattice M_I^G is irreducible [46]. Moreover, by Frobenius reciprocity [C-R1; 82]

$$End_{SG}(M_I^G) \simeq Hom_{SI}(M, res_I(M_I^G)),$$

where the restriction of the induced module M_I^G to I,

$$res_I(M_I^G) \simeq_{SI} \oplus_{i=1}^t {}^{g_i}M$$

is the direct sum of the conjugate modules ${}^{g_i}M$ for coset representatives $\{g_i\}$ of I in G. According to the definition, ${}^{g_i}M \simeq_{SI} M$ if and only if $g_i = 1$. Hence

---

46   This also follows from ordinary Clifford theory

$End_{SG}(M_I^G) = S$, since $End_{SI}(M) = S$ – recall, that $\phi(g)$ is S-linear. On the other hand,

$$End_{R_0 G}(M_I^G) \simeq Hom_{R_0 I}(M, res_I(M_I^G)), \text{ with}$$

$$res_I(M_I^G) \simeq_{R_0 I} \oplus_{i=1}^{t} {}^{g_i}M$$

the direct sum of the conjugate modules ${}^{g_i}M$ for coset representatives $\{g_i\}$ of I in G. According to the definition, ${}^{g_i}M \simeq_{R_0 I} M$ for all $1 \leq i \leq t$. Hence $\Lambda := End_{R_0 G}(M_I^G) = S^t$.

We recall [C-R1; 82, (8.16)], that for $R_0 G$-lattices X and Y, we have

$$S \otimes_{R_0} Hom_{R_0 G}(X, Y) \simeq Hom_{SG}(S \otimes_{R_0} X, S \otimes_{R_0} Y).$$

Thus

$$S \otimes_{R_0} S^t \simeq End_{SG}((S \otimes_{R_0} M)_I^G).$$

XIII.14. Claim.

$$(S \otimes_{R_0} M)_I^G \simeq \sum_{1}^{t} M_I^G \text{ as } SN- \text{ modules,}$$

and hence $End_{SG}((S \otimes_{R_0} M)_I^G) \simeq (S)_t$, and $S \otimes_{R_0} \Lambda \simeq (S)_t$.

PROOF OF CLAIM XIII.14: We have seen in XIII.12, that the group conjugate module ${}^{g_i}M$ is SN-isomorphic to the Galois conjugate module ${}^{\rho(g_i)}M$. Since $S \otimes_{R_0} M$ is RN-isomorphic to the direct sum of the Galois conjugate modules $\{M_i\}_{1 \leq i \leq t}$, we conclude,

$$S \otimes_{R_0} M \simeq_{SN} M_I^G.$$

In particular, $(M_i)_I^G \simeq_{SG} M_I^G$. Thus $S \otimes M_I^G \simeq \sum_1^t M_I^G$, and Claim XIII.14 is proved. q.e.d.   □

Recall, $\Lambda = End_{RG}(M_I^G)$, and $S \otimes_{R_0} \Lambda \simeq (S)_t$, where S is an unramified extension of R_0. Thus $S \otimes_{R_0} rad(\Lambda) = rad(S \otimes_{R_0} \Lambda)$. Since $rad(S \otimes_{R_0} \Lambda) \simeq S \otimes_{R_0} \Lambda$ as modules $S \otimes_{R_0} \Lambda$, ($S \otimes_{R_0} \Lambda$ being isomorphic to the full t by t matrix ring over S), we can invoke the Noether – Deuring theorem [C-R1; 82] to conclude $rad(\Lambda) \simeq \Lambda$. But then Λ is hereditary [Re; 75, (39.1)][47]. If now P is an indecomposable projective left Λ-module, then $P \simeq rad(\Lambda)$, since this holds when extended to S. Thus Λ is a maximal order with $S \otimes_{R_0} (\Lambda/rad(\Lambda)) \simeq (S/rad(S))_t$, and we conclude, that Λ decomposes into t isomorphic left modules ([Re; 75, (18.4)]).

Moreover, R_0 lies in the centre of Λ, and $dim_{R_0}(\Lambda) = dim_{R_0}(S^{(t)}) = t^2$. Hence the only possibility for Λ is $\Lambda \simeq (R_0)_t$. Since $\Lambda = End_{RG}(M_I^G)$, this shows, that as RG-module, M_I^G decomposes into t RG-modules, say

---

47   every Λ-lattice is projective

$X_i, 1 \leq i \leq t$; however, as RI–module, $M_I^G \simeq M^{(t)}$, and we conclude, that M_I^G is the direct sum of t isomorphic RG–modules, which when restricted to I are isomorphic to M.

Thus M extends to an RG–module. It is also an R_0G–module. This completes the proof of XIII.13   □

XIII.15. Remark. We shall now treat the situations "splitting field" or "non splitting field" separately, and we shall first deal with the split situation, which is the classical set up for Clifford theory [C-R1; 82, (11.1)] :

M is as above an irreducible SN–lattice with $End_{SN}(M) = S$ and inertia group $I_S(M)$. Moreover, $(|I_S(M) : N|, |N|) = 1$, and so M extends as S–module to its inertia group $I_S(M)$ (see XIII.7).

XIII.16. Theorem. *[Mo; 58, Ga; 79] With the above assumptions, the group ring SG has a ring direct summand of the form*

$$B_M = Mat_{|G:I_S(M)|}(SI_S(M)/N \otimes_S \Lambda_M) \simeq Mat_{|G:I_S(M)|\cdot n}(SI_S(M)/N) ,$$

where $\Lambda_M \simeq End_S(M) = Mat_n(S)$ is the two sided ideal of SN corresponding to M.
In particular,

$$SG = \prod Mat_{|G:I_S(M)|}(S(I_S(M)/N) \otimes_S \Lambda_M),$$

where the product is taken over representatives of indecomposable SN–lattices, which are not G–conjugate. Thus SG is Morita equivalent to $\prod S(I_S(M)/N)$.

PROOF OF THEOREM XIII.16: **Case 1:** First let $G = I_S(M)$.

Let e_M be the central primitive idempotent of SN, which corresponds to the indecomposable SN–lattice M; i.e. $SN \cdot e_M = End_S(M) = (S)_n$. According to the definition of the inertia group, e_M is also a central idempotent in SG, not necessarily primitive. In fact, the conjugate idempotents ${}^g e_M$ are central idempotents in RN, which are all equivalent to e_M, since G is the inertia group of M.

We shall now focus our attention first on the ring direct summand $SG \cdot e_M$ of SG.

We claim that the module–the ordinary tensor product of SG–modules–

$$P_M = SG/N \otimes_S M \text{ is projective for } SG.$$

For this it suffices to show, that the restriction to a Sylow p–subgroup P is projective; however, $(|G/N|, |N|) = 1$ and $|N| \cdot S = S$. Hence G is a semi direct product of G/N and N, and G/N contains a Sylow p–subgroup of G. Consequently, SG/N is SG–projective. But then the tensor product $SG/N \otimes_S M$ is projective [Ben; 84, Lemma 2.1.5] as SG–module, it is even a cyclic SG–module, generated by $1 \otimes m_0, 0 \neq m_0 \in M$.

We now compute the ring of SG–endomorphisms of P_M:

$$
\begin{aligned}
End_{SG}(SG/N \otimes_S M) &= (End_{SN}(SG/N \otimes_S M))^{G/N} \\
&= (End_{SN}(SG/N) \otimes_S End_{SN}(M))^{G/N} \\
&= (SG/N \otimes_S S),
\end{aligned}
$$

(N acts trivially on SG/N) where $X^{G/N}$ are the G/N fixed points on X. G/N acts on SG/N from the left and on S it acts trivially. Hence

$$
End_{SG}(SG/N \otimes_S M) \simeq SG/N.
$$

We now have to compute the multiplicity of P_M in SG. The group ring SG is a twisted tensor product, $SG \simeq SG/N \otimes_S SN$, and since e_M is a central idempotent of SG, the two sided $SG \simeq SG/N \otimes_S SN$–module $SG/N \otimes_S \Lambda_M$ is a ring direct summand of SG, which is as left module isomorphic to $P_M^{(n)}$. Thus the group ring SG has a ring direct summand isomorphic to $(SG/N)_n$, which is the statement of Theorem XIII.16 in case $I_S(M) = G$.

Case 2: $N \leq I_S(M) < G$.

Let

$$
G = \cup_{1 \leq i \leq s} \; g_i \cdot I_S(M)
$$

be a system of left coset representatives of $I_S(M)$ in G. e_M is the central primitive idempotent in SN corresponding to M. Since we are outside of the inertia group of M, the idempotents $\{{}^{g_i}e_M\}_{1 \leq i \leq s}$ are different primitive orthogonal central idempotents of SN – note that N is normal.

We put $e_i = {}^{g_i}e_M$, $1 \leq i \leq s$. Then $e = \sum_{1 \leq i \leq n} e_i$ is a central idempotent in SG.

We recall from above that e_M is also a central idempotent in $SI_S(M)$. G acts on the idempotents e_i as follows:

If $g_j g_i = g_k x$, $x \in I_S(M)$, then

$$
g_j e_i = e_k x,
$$

and hence if L is an $SI_S(M)$–module on which e_M acts as identity, then $\oplus e_i L$ is the induced module $L \uparrow_{I_S(M)}^G$.

XIII.17. Claim. SG contains a ring direct summand, which is Morita equivalent to

$$
\Gamma := SI_S(M)/N \otimes_S End_S(M),
$$

which in turn is Morita equivalent to $SI_S(M)/N$.

PROOF OF CLAIM XIII.17: The idempotents $\{e_i\}_{1 \leq i \leq s}$ are surely orthogonal and their sum $e := \sum_{1 \leq i \leq s} e_i$ is a central idempotent in SG. Thus we have a

Pierce decomposition

$$
B := \begin{pmatrix}
e_1 \cdot SG \cdot e_1 & \dots & e_1 \cdot SG \cdot e_j & \dots & e_1 \cdot SG \cdot e_s \\
\dots & \dots & \dots & \dots & \dots \\
e_i \cdot SG \cdot e_1 & \dots & e_i \cdot SG \cdot e_j & \dots & e_i \cdot SG \cdot e_s \\
\dots & \dots & \dots & \dots & \dots \\
e_s \cdot SG \cdot e_1 & \dots & e_s \cdot SG \cdot e_j & \dots & e_s \cdot SG \cdot e_s
\end{pmatrix}.
$$

On the other hand, the module induced to G from the projective $I_S(M)$–module $P_M = SI_S(M)/N \otimes_S End_S(M)$ is

(53)
$$
\begin{aligned}
P_0 &:= (SI_S(M)/N \otimes_S End_S(M))^G_{I_S(M)} \\
&\simeq SG \otimes_{SI_S(M)} (SI_S(M)/N \otimes_S End_S(M)) \\
&\simeq \oplus_{1 \le i \le s} e_i \cdot SI_S(M)/N \otimes_S End_S(M).
\end{aligned}
$$

This latter isomorphism is a consequence of the discussion preceding XIII.17.

Moreover, e acts as identity. Surely the modules $P_0 \cdot e_j$ are isomorphic to P_0, $1 \le j \le s$, and hence their direct sum is a two sided direct summand of SG, and it is $e \cdot SG$. So the claim is proved.   □

q.e.d.Theorem XIII.16   □

This finishes the argument in the splitting situation. We now turn to the general set up:

XIII.18. Recall.

(a) M is an irreducible RN–lattice, with $End_{RN}(M) = S$, and hence can be viewed as an SN–module.

(b) $I(M)$ is the inertia group of M as RN–module, $I_S(M)$ is the inertia group of M as SN-module. $I_S(M)$ is a normal subgroup in $I(M)$ with cyclic quotient $T(M) = I(M)/I_S(M)$, which injects into $Gal(S/R)$ (see XIII.8). $T(M)$ has fixed ring R_0 in S.

(c) We assume as above that $(|I_S(M) : N|, |N|) = 1$. By XIII.13 M extends to an $SI(M)$-module, with $End_{SI(M)}(M) = S$, it extends to an $RI(M)$–module with $End_{RI(M)}(M) = R_0$. Moreover, the induced $SI(M)$–module $M^{I(M)}_{I_S(M)}$ is irreducible; as $SI(M)$–module $M^{I(M)}_{I_S(M)}$ is the direct sum of the Galois conjugates (XIII.12), whereas the induced $RI(M)$ (even $R_0I(M)$)–module $M^{I(M)}_{I_S(M)}$ decomposes in $t = |T(M)|$ copies of M (see XIII.13).

(d) By Theorem XIII.16 the group ring $SI(M)$ contains a ring direct summand

(54) $\Delta_S(M) := (SI_S(M)/N \otimes_S (S)_n)_t \simeq (SI_S(M)/N)_{n \cdot t}$

(e) Moreover, we have seen in XIII.16, that the $SI(M)$–module $(SI_S(M)/N \otimes_S M)^{I(M)}_{I_S(M)}$ and hence also $(SI_S(M)/N \otimes End_R(M))^{I(M)}_{I_S(M)}$

are projective over $SI(M)$. But since

$$S \otimes_R ((RI_S(M)/N \otimes_R M)_{I_S(M)}^{I(M)}) \simeq (SI_S(M)/N \otimes_S M)_{I_S(M)}^{I(M)},$$

we conclude that $(RI_S(M)/N \otimes_R M)_{I_S(M)}^{I(M)}$ is a projective $RI(M)$–module.

XIII.19. Remark. In general I can not say any more but that RG contains a ring direct summand, which is a full matrix ring over

$$End_{RI(M)}((RI_S(M)/N \otimes_R M)_{I_S(M)}^{I(M)}).$$

However, the situation is much more transparent, if we *assume*, that the group extension $I(M)/N$ over $I_S(M)/N$ is split.

That this need not always be so shows an example in XIV3.2, which is a slight modification of example 3.1.

In the non split situation, the 2–cocycle $\psi \in H^2(T(M), S^*)$ probably will play an important role.

XIII.20. Assumption. The exact sequence of groups

$$1 \longrightarrow I_S(M) \longrightarrow I(M) \longrightarrow I(M)/I_S(M) \longrightarrow 1$$

is split. Let $T_0(M)$ be a subgroup of $I(M)$, which is mapped isomorphically onto $T(M) := I(M)/I_S(M)$.

XIII.21. Lemma. *The $R_0T_0(M)$–module S° which is S with $T_0(M)$ acting as Galois automorphism is as left module $R_0T_0(M)$–isomorphic to $R_0T_0(M)$. If $R_0(I_S(M)^\circ)$ is the free left $R_0I_S(M)$–module of rank one with $T_0(M)$ acting by conjugation, then the $R_0I(M)$–module $S^\circ \otimes_{R_0} R_0(I_S(M)^\circ)$, which we shall denote by $SI_S(M)^\circ$, is as left $R_0I(M)$–module isomorphic to $R_0I(M)$.*

PROOF OF LEMMA XIII.21: The fact, that S° is free as $R_0Gal(S/R_0) \simeq R_0T_0(M)$–module is a result of David Hilbert [Que; 80, p.219] [Hil; 1897], since S is unramified over R_0. The remaining statement is a general fact about semi direct products: Let $A \rtimes B$ be a semi direct product of groups, with B acting on A by conjugation, then $R(A^\circ) \otimes_R RB$ with B acting on $R(A^\circ)$ by conjugation is isomorphic to RG.

q.e.d. Lemma XIII.21   □

N° is the group N but $T(M)$ acting via conjugation.

XIII.22. Lemma. *The $RI(M)$–module $RI_S(M)/N^\circ \otimes_R M$ where $I(M) = I_S(M) \rtimes T_0(M)$ and $T_0(M)$ acts on $RI_S(M)/N^\circ$ by conjugation, is a projective $RI(M)$–module.*

PROOF OF LEMMA XIII.22: Since the Sylow p–subgroup of $I(M)$ injects into the Sylow p–subgroup of $I(M)/N$, the module $SI_S(M)/N^\circ$ is a projective

$R_0I(M)$–module by XIII.21, and hence the tensor product $SI_S(M)/N^\circ \otimes_{R_0} M$
is a projective $R_0I(M)$–module, and hence also a projective $RI(M)$–module.
However, we have the following chain of isomorphisms

$$
\begin{aligned}
SI_S(M)/N^\circ \otimes_R M \quad &\simeq_{R_0I(M)} \quad SI_S(M)/N^\circ \otimes_R R_0) \otimes_{R_0} M \\
&\simeq_{R_0I(M)} \quad (R_0 \otimes_R S)I_S(M)/N^\circ \otimes_{R_0} M \\
&\simeq_{R_0I(M)} \quad S^{|R_0:R|}I_S(M)/N^\circ \otimes_{R_0} M \\
&\simeq_{RI(M)} \quad \oplus RI(M)/N^\circ \otimes_{R_0} M.
\end{aligned}
$$

The third isomorphism holds since S is unramified over R. Thus the statement
of the lemma follows.

q.e.d. Lemma XIII.22   □

The tensorproduct $X := R(I_S(M)/N^\circ) \otimes_R S$ is then a $T_0(M)$–module,
where $T_0(M)$ acts as Galois automorphisms via $T(M)$ on S and by conjugation
on $I_S(M)$. Then $End_{T_0(M)}(X) \simeq H^0(T_0(M), R(I_S(M)/N^\circ) \otimes_R S)$ is the ring
of fixed points in $R(I_S(M)/N^\circ) \otimes_R S$ under the diagonal action of $T_0(M)$.

XIII.23. Lemma. *The $RI(M)$–module $Y := RI_S(M)/N^\circ \otimes_R M$ has*

$$End_{RI(M)}(Y) = H^0(T_0(M), R(I_S(M)/N^\circ) \otimes_R S).$$

PROOF OF LEMMA XIII.23: We compute the endomorphism ring of Y as

$$
\begin{aligned}
End_{RI(M)}(Y) &= End_{RI(M)}(RI_S(M)/N^\circ \otimes_R M) \\
&= (End_{RI_S(M)}(RI_S(M)/N^\circ \otimes_R M))^{T_0(M)} \\
&= (RI_S(M)/N^\circ \otimes_R S^\circ)^{T_0(M)} \\
&= H^0(T_0(M), RI_S(M)/N^\circ \otimes_R S^\circ).
\end{aligned}
$$

q. e. d. Lemma XIII.23   □

We now can state the main theorem in this section:

Theorem. *Assume that N is a normal subgroup of the finite group G, with
$|N| \cdot R = R$ and $(|N|, |G:N|) = 1$. Let M be an irreducible RN–lattice with
$End_{RN}(M) = S$ a finite unramified extension of R; let $n = dim_S(M)$, and
denote by $I(M)$ and $I_S(M)$ be the inertia groups of M as RN–module and
SN–module resp. Then $I_S(M) \trianglelefteq I(M)$ with quotient $T(M)$, which acts as
group of R–Galois automorphisms on S. Assume furthermore, that $I_S(M)$ has
a complement $T_0(M)$ in $I(M)$. Denote by $SI_S(M)/N^\circ$ the $RI(M)$–module,
where $I_S(M)$ acts by left multiplication and $T_0(M)$ acts by conjugation. S° is
the $RT_0(M)$–module, where $T_0(M)$ acts via $T(M)$ as Galois automorphism.*

Then the group ring RG contains a ring direct summand of the form

$$B := Mat_{|G:I_S(M)|\cdot n}(H^0(T_0(M), RI_S(M)/N^\circ \otimes_R S^\circ)).$$

PROOF OF THEOREM XIII.2: Putting together the statements from the lemmata XIII.21, XIII.22, XIII.23 we conclude that the group ring $RI(M)$ contains a ring direct summand of the form

$$B_{I(M)} := Mat_{|I(M):I_S(M)| \cdot n}(H^0(T_0(M), RI_S(M)/N^\circ \otimes_R S^\circ)).$$

Passing from $I(M)$ to G is done exactly as in the proof of Theorem XIII.16, where one passed from the inertia group of M, $I_S(M)$, to G. This process gives a full $|G : I(M)|-$ matrix ring over $B_{I(M)}$.

q. e. d. Theorem XIII.2   □

XIV Examples

One might be tempted by the result, that for a solvable group G with $O_{p'}(G) = 1$ group bases are p–adically conjugate, to think, that finite p–subgroups of $V(\hat{\mathbb{Z}}_p G)$ are conjugate to subgroups of G. We present next an example, which shows, that this is not so.

§ 1   Extending the cyclic group of order 6

Let $S_3 = < a, b : a^3, b^2, {}^b a = a^{-1} >$ be the symmetric group on three letters, let $C_3 = < c : c^3 = 1 >$ be the cyclic group of order 3, and put $G = C_3 \times S_3$. Note that G is an extension of C_6. Then $e_1 = (b+1)/2$ and $e_2 = (1-b)/2$ are orthogonal idempotents in $\hat{\mathbb{Z}}_3 G$ with $e_1 + e_2 = 1$. Moreover the group ring $\hat{\mathbb{Z}}_3 G = \hat{\mathbb{Z}}_3 C_3 \otimes_{\hat{\mathbb{Z}}_3} \hat{\mathbb{Z}}_3 S_3$ is a tensor product of rings.

We now consider the element $u = (c \otimes e_1) + (c^{-1} \otimes e_2)$. Then $u \in \hat{\mathbb{Z}}_3 G$ is a unit of order 3, which is also augmented. u cannot be conjugate to an element of G, not even in $\hat{\mathbb{Q}}_3 G$, since u is not even a group element in the commutative quotient $\hat{\mathbb{Z}}_3 (C_3 \times < b >)$. Note also, that u is not central, since $a \cdot u \ne u \cdot a$.

We also point out, that u cannot be the localization of a global unit of augmentation one, since $\mathbb{Z}(C_3 \times < b >)$ has only trivial units of augmentation one.

This example has another noteworthy property: The group $< u >$ is isomorphic to a subgroup – namely $< a >$ – of G which is a $p = 3$–group such that the centralizer is a p–group. This is of importance with respect to chapter VI Theorem VI.1 and a possible generalization in the direction of 1.1.

We may construct as well a unit of this group ring of order 3 with coefficient of 1 not being zero. Indeed $c \otimes e_2 + e_1$ is such an element. This example also works if we reduce modulo $I(< a >)G$ to the group ring of the cyclic group of order 6.

§ 2   Sylow subgroups in the principal block

The following example appeared in a discussion with W. Kimmerle:

Since the dihedral group of order 8 embeds into the group ring $\hat{\mathbb{Z}}_3 S_3$, we may construct two embeddings of a 2–Sylow subgroup of the symmetric group S_4 of order 24, into the group ring of the symmetric group of order 24 over the 2 – adic integers, which are not conjugate in the group ring:

More precisely

$$S_4 = (C_2 \times C_2) \rtimes (C_3 \rtimes C_2).$$

Then since D_8 embeds into

$$\hat{\mathbb{Z}}_2 S_3 = (\hat{\mathbb{Z}}_2)_2 \times \hat{\mathbb{Z}}_2 C_2$$

via the faithful representation of D_8 in $(\hat{\mathbb{Z}}_2)_2$, which in turn embeds into $\hat{\mathbb{Z}}_2 S_4$. However, on the other hand, the 2–Sylow subgroups of S_4 are dihedral groups of order 8 and from there one gets another embedding of the dihedral group of order 8, which is not conjugate to the embedding above, not even over $\hat{\mathbb{Q}}_2 G$.

2.1. Proposition. *In the group ring $\hat{\mathbb{Z}}_2 S_4$ of the symmetric group on 4 letters there are two non conjugate subgroups of order 8. One of these subgroups consists of linear dependent elements, the other does not. This happens though $O_{2'}(S_4) = 1$.*

§3   Examples concerning Clifford's theory

3.1. Example. The structure of the inertia groups and also the structure of blocks is quite different in case $R = \hat{\mathbb{Z}}_p$ and in the splitting situation, as shows the following *example*, which arose in a discussion with Gerhard Hiss:
Let

$$G = < a, b, c \mid a^4, b^3, c^2, [a, b],\ {}^c a = a^{-1},\ {}^c b = b^{-1} >,$$

and put $S = \hat{\mathbb{Z}}_3[\zeta]$, where ζ is a primitive 3-rd root of unity. We shall be dealing with the non principal block B of SG.

Case 1: We consider the group ring SG. Let M_0 be one of the *two* non isomorphic non trivial irreducible $S < b >$ –modules. (Note that these modules are isomorphic, when considered as $\hat{\mathbb{Z}}_3 < b >$–modules.) The inertia group of M_0 is then $< a > \times < b >$, and hence Clifford theory (cf. below) shows:

$$B = (M_0 < a >)_2 \simeq (S < a >)_2.$$

Here the group ring $S < a >$ is a direct product

$$S < a > \simeq S[i] \times S \times S.$$

Case 2: We consider the group ring $\hat{\mathbb{Z}}_3 G$. Here, M_0 is *the* non trivial irreducible $\hat{\mathbb{Z}}_3 < b >$ –module. The inertia group is then G itself. An easy calculation (cf. below) shows that

$$B = (\Lambda)_2,$$

where Λ is the pullback

$$
\begin{array}{ccc}
\Lambda & \to & \hat{\mathbb{Z}}_3 \\
\downarrow & & \downarrow \\
S & \to & \mathbb{Z}/3\mathbb{Z}
\end{array}
$$

with S is the fixed ring of $\hat{\mathbb{Z}}_3[\zeta, i]$ under the diagonal involution. It is totally ramified, but not isomorphic to $R[i]$.

This example shows at the same time, that the theorem of Puig for nilpotent blocks [Pu; 81], which in our case states, that in the splitting case B is Morita equivalent to the group ring of the defect group $< a >$ does not hold in the non splitting situation.

3.2. Example. We next shall give an example that the splitting of $I(M)$ over $I_S(M)$ is not automatic:

Let $H = I\!\!F_7 \rtimes I\!\!F_7^*$ be the affine group of the line over $I\!\!F_7$. Then $C_3 \leq I\!\!F_7^*$ leads to $K := I\!\!F_7 \rtimes C_3$. We form the pullback along the central extension

$$1 \longrightarrow C_3 \longrightarrow C_9 \longrightarrow C_3 \longrightarrow 1$$

and get the group $G = I\!\!F_7 \rtimes C_9$. Let M be a faithful irreducible $\hat{\mathbb{Z}}_3 C_7$ lattice. Since its dimension over $\hat{\mathbb{Z}}_3$ is 6, it is unique. We choose for Clifford's Theorem $N = I\!\!F_7$ and $R = \hat{\mathbb{Z}}_3$. Therefore $S = End_{\hat{\mathbb{Z}}_3 C_7}(M) = \hat{\mathbb{Z}}_3[\zeta_7]$ with ζ_7 being a primitive 7th root of unity. The exact sequences

$$0 \longrightarrow I(N) \uparrow^G \longrightarrow RG \longrightarrow RC_9 \longrightarrow 0$$

and

$$0 \longrightarrow I(N) \uparrow^{G/C_3} \longrightarrow RG/C_3 \longrightarrow RC_3 \longrightarrow 0$$

are two side split. Since $I(N) \uparrow^{G/C_3}$ is a block of defect 0, M lying inside, $I(N) \uparrow^G$ is a block of defect at most 1. Hence $I_S(M) = I\!\!F_7 \times C_3$, however $I(M) = G$, since M has no conjugates over R. Therefore $I_S(M)$ has no complement in $I(M)$.

References

[Al; 86]Alperin J.: Local Representation Theory, Cambridge University Press, 1986.

[A-M; 69]Atyah M., I.G. Macdonald: Introduction to commutative algebra, Oxford 1969.

[Ar; 84]Artin E.: The orders of the classical groups, Comm. Pure Appl. Math. 8 (1984), 446–460.

[Asch; 87]Aschbacher M.: Finite Group Theory, Cambridge Studies in Math. 1987.

[Atl; 85]Conway J., R. Curtis, S. Norton, R. Parker, R. Wilson: Atlas of Finite Groups, Oxford University Press, Oxford 1985.

[Ba; 64]Bass H.: The stable structure of quite general linear groups, J. AMS 70 (1964), 429–433.

[B-M-S; 67]Bass H., J. Milnor, J.P. Serre: Solution of the congruence subgroup problem for Sl_n ($n \geq 3$) and Sp_{2n} ($n \geq 2$); Publ. IHES 33 (1967), 59–137.

[Ben; 84]Benson D.: Modular Representations, New Trends and Methods, Springer Lecture Notes in Mathematics 1081 (1984).

[Ber; 53]Berman S.D.: On certain properties of integral group rings; (Russian) Dokl. Akad. Nauk SSSR (N.S.) 91 (1953), 7–9.

[Ber; 55]Berman S.D.: On the equation $x^m = 1$ in an integral group ring, Ukrain. Mat. Zh. 7 (1955), 253–261.

[Bes; 89]Bessenrodt C.: The isomorphism type of an abelian defect group of a block is determined by its modules. J.London Math.Soc.(2) 39 (1989), 61-66.

[Bo; 60]Bourbaki N.: Éléments de mathématique première partie, topologie générale, chapitre 3 groupes topologiques (théorie élémentaire); Actualités sci. et ind. 1143, 1960.

[Bra; 63]Brauer R.: Representations of finite groups. Lectures on modern mathematics, Vol.I, 133-175, Wiley New York, 1963. reproduced in Vol.II of: Richard Brauer: collected papers. MIT press, Cambridge MA, 1980.

[Br; 87]Brown K.S.: Cohomology of Groups, Springer Graduate Texts in Mathematics 87, (1987).

[C-S-W; 81]Cliff G.H., S.K. Sehgal, A.R. Weiss: Units of integral group rings of metabelian groups; J. of Algebra 73 (1981), 167–185.

[C-H; 80]Camina A., M. Herzog: Character tables determine abelian Sylow 2 – subgroups, Proc. AMS 80 (1980) 533–535.

[C-E; 56]Cartan H., S. Eilenberg: Homological Algebra; Princeton University Press, 1956.

[Col; 64]Coleman D.: On the modular group ring of a p–group, Proc. AMS, 15 (1964), 511–514.

[Con; 72]Conlon S.B.: A basis for monomial algebras; J. of Alg. 20 (1972), 396–415.

[C-R; 62]Curtis C.W. I. Reiner: Representation Theory of finite Groups and Associative Algebras, John Wiley (1962).

[C-R1; 82]Curtis C.W., I. Reiner: Methods of Representation Theory, Vol. 1, John Wiley Interscience (1982).

[C-R2; 87]Curtis C.W., I. Reiner: Methods of Representation Theory, Vol. 2, John Wiley Interscience (1987).

[Da; 64,1]Dade E.: Deux groupes finis ayant la même algèbre de groupe sur tout corps, Mathematische Zeitschrift 119 (1964), 345–348.

[Da; 64,2]Dade E.: Answer to a question of R. Brauer; J. of Algebra, 1 (1964), 1–4.

[De; 77]Dennis K.: The structure of the unit group of a group ring; Proceedings Ring Theory Conference, University of Oklahoma (1976), 103–130.

[Di; 01]Dickson L.E.: Linear Groups, B.G. Teubner, Leipzig 1901; republished by Dover, N.Y. 1958.

[Er; 90]Erdmann K.: Blocks of tame representation type and related algebras; Lecture Notes in Math. 1428, Springer, Berlin 1990.

[Fe; 80]Feit W.: Some consequences of the classification of finite simple groups. The Santa Cruz conference on finite groups, 175-181, Proc.Sympos.Pure Math.37, Amer.Math.Soc.,Providence, 1980.

[Fr; 73]Fröhlich A.: The Picard groups of noncommutative rings, in particular of orders, Trans. Amer. Math. Soc., 180 (1973), 1–46.

[F-R-U; 74]Fröhlich A., I. Reiner, S. Ullom:Class groups and Picard groups of orders, Proc. London Math. Soc. (3) 29 (1974), 405–434.

[F-K-W; 74]Fröhlich A., M.E. Keating, S.M.J. Wilson : The class group of quaternion and dihedral 2–groups, Mathematika 21 (1974), 64–71.

[Ga; 79]Gallagher P.X.: Invariants for finite groups, Adv. Math. 34 (1979), 46–57.

[Gl; 65]Glauberman G. as quoted in [Pa; 65].

[G-L; 83]Gorenstein D., R. Lyons: The local structure of finite groups of characteristic 2 type. Memoirs AMS, Vol. 42, No. 276, Providence R.I. 1983.

[Gr; 67]Gruenberg K.: Profinite Groups; Cassels, Fröhlich, Algebraic Number Theory, Proceedings of an International Conference held at Brighton, Academic Press (1967).

[G-R; 88]Gustafson W., K.W. Roggenkamp: A Mayer – Vietoris sequence for Picard groups with applications to integral group rings of dihedral and quaternion groups, Ill. J. Math. (1988) I. Reiner memorial volume, 375–406.

[H-L; 90]Hiss G., K.Lux: Brauer Trees of sporadic groups, Oxford University Press 1990

[Has; 49]Hasse H.: Zahlentheorie, Akademie–Verlag, Berlin 1949.

[Har; 77]Hartshorne R.: Algebraic Geometry, Graduate Texts in Mathematics 52, Springer (1977).

[Hig; 39]Higman G.: Units in group rings, D. phil. theses, Oxford Univ. (1939).

[Hil; 1897]Hilbert D.: Die Theorie der algebraischen Zahlkörper, Gesammelte Abhandlungen Band 1, Springer 1970.

[H-S; 70]Hilton D., U. Stammbach: A Course in Homological Algebra, Springer, 1970.

[Hu; 67]Huppert B., Endliche Gruppen I, Springer Verlag 1967.

[H-B2; 82]Huppert B., N. Blackburn, Finite Groups II, Springer 1982.

[H-B3; 82]Huppert B., N. Blackburn, Finite Groups III, Springer 1982.

[H-P; 72]Hughes I., K.E. Pearson: The group of units of the integral group ring $\mathbb{Z}S_3$, Can. Math. Bull 15 (1972), 529–534.

[Is; 76]Isaacs M.: Character Theory of Finite Groups. Academic Press, N.Y. 1976.

[J-M; 87]Jackowski S., Z. Marciniak: Group automorphisms inducing the identity map on cohomology; J. of Pure and Appl. Algebra 44 (1987), 241–250.

[Ja; 69]Jackson D.A.: The group of units of the integral group rings of finite metabelian and finite nilpotent groups, Quart. J. Math. Oxford (2) 20 (1969), 313–319.

[Ka; 74]Kaplansky I.: Commutative Rings, Queen Mary College Mathematics Notes. 1974

[Ki; 91]Kimmerle W.: Beiträge zur ganzzahligen Darstellungstheorie endlicher Gruppen, Bayreuther Math. Schr. Heft 36 (1991).

[K-L-S-T; 90]Kimmerle W., R. Lyons, R. Sandling, D. Teague: Composition factors from the group ring and Artin's theorem on orders of simple groups; Proceedings LMS (3) 60 (1990), 89–122.

[K-R; 91]Kimmerle W., K. Roggenkamp: A Sylowlike Theorem for Integral Group Rings of Finite Solvable Groups, appears in Arch. d. Math.

[K-S; 91]Kimmerle W., R. Sandling: A group theoretic and group ring theoretic dtermination of certain Sylow and Hall subgroups and the resolution of a question of Brauer; preprint.

[Kl; 91]Klingler L.: Construction of a counterexample to a conjecture of Zassenhaus, Comm. Alg., 19 (8), (1991), 2303–2330.

[Mic; 86]Michler G.: Brauer's conjectures and the classification of finite simple groups. Groups and orders (Ottawa 1984), Lecture Notes in Math., Springer, Berlin 1986, 129 – 142

[Mil; 71]Milnor J.: Introduction to algebraic K–theorie, Ann. of Math. Studies 72 1971, Princeton N.J.

[Mo; 58]Morita K.: Duality for modules and its applications to the theory of rings with minimum condition; Sci. Rep. Tokyo Kyoiku Daigaku, Section A, 6, no. 150, (1958), 83–142.

[Pa; 65]Passman D.S.: Isomorphic Groups and Group Rings, Pacific Journal of Mathematics (2) 35 (1965), 561–583.

[Pa; 77]Passman D.S.: Algebraic structure of group rings; Interscience, N.Y. 1977.

[Pe; 76]Peterson G.: Automorphisms of the integral group rings of S_n, Proceedings AMS Vol. 59, No.1, (1976), 14–18.

[Pl; 91]Plesken W.: Some applications of representation theory; in Representation Theory of Finite Groups and Finite–Dimensional Algebras, ed. by G.O. Michler and C.M. Ringel, Birkhäuser Progress in Math. Vol. 95, 477–496.

[Pu; 81]Puig L.: Pointed groups and construction of characters, Math. Z. 176 (1981), 165–292.

[Qui; 71]Quillen D.: The spectrum of an equivariant cohomology ring: I and II, Annals of Math. 94 (1971), 549–602.

[Que; 80]Queyrut J.: S–groupes de Grothendiecket structure galoisienne des anneaux d'entiers, Springer Lecture Notes in Math. 882, pp.219–239, 1980.

[Re; 75]Reiner I.: Maximal Orders, Academic Press (1975).

[R-R; 79]Reiner I., K.W. Roggenkamp: Integral Representations, Springer Lecture Notes in Mathematics 744, (1979).

[R-U; 74]Reiner I., S. Ullom: Mayer Vietoris sequence for class groups; J. of Alg. 31 (1974), 305–342.

[Ro; 72 1]Roggenkamp K. W.: Integral Representations of Finite Groups, Presses Universités Montreal, (1972).

[Ro; 72 2]Roggenkamp K. W.: An extension of the Noether–Deuring Theorem, Proc. AMS 31 (1972), 423–426.

[Ro; 80]Roggenkamp K. W.: Grouprings of Metabelian Groups and Extension Categories; Canadian Journal of Mathematics XXXII, No.2 (1980), 449–459.

[Ro; 81]Roggenkamp K.W.: Units in integral metabelian group rings I, Jackson's unit theorem revisited; Quart. J. of Math. Oxfd. (2), 32, No. 126 (1981), 209–224.

[Ro; 92]Roggenkamp K.W.: Subgroup rigidity of p–adic group rings (Weiss' arguments revisited); to appear J. Lon. Math. Soc. (1992).

[R-S; 83]Roggenkamp K.W., L.L. Scott: Units in Metabelian Group Rings: Non–Splitting Examples for Normalized Units; J. of Pure and Appl. Algebra 27 (1983), 299–314.

[R-S; 85]Roggenkamp K.W., L.L. Scott: Units in Group Rings: Splittings and the Isomorphism Problem; J. of Algebra, Vol. 96, No. 2 (1985), 397–417.

[R-S; 86]Roggenkamp K. W., L.L.Scott: The Isomorphism Theorem for Integral Group Rings of Nilpotent by Abelian Groups, manuscript, (1986).

[R-S; 87 1]Roggenkamp K. W., L.L. Scott: Isomorphisms of p-adic group rings;

Annals of Mathematics, 126 (1987), 593-647.

[R-S; 87 2]Roggenkamp K. W., L.L. Scott: A strong answer to the isomorphism problem for finite p-solvable groups with a normal p-subgroup containing its centralizer, manuscript, (1987).

[R-S; 87 3]Roggenkamp K. W., L.L. Scott: On a Conjecture of Zassenhaus, preprint (1987).

[R-Z; 90]Roggenkamp K. W., A. Zimmermann: On the Isomorphism Problem for Integral Group Rings of Finite Groups, preprint, (1990).

[Sah; 68]Sah C.F.: Automorphisms of finite groups, J. Alg. 10 (1968), 47–68.

[San; 74]Sandling R.: Group rings of circle and unit groups; Math. Z. 140 (1974), 195–202.

[San; 85]Sandling R.: The isomorphism problem for group rings: a survey, in: Lecture Notes in Math., 1142, Springer Berlin (1985), pp. 256–288.

[Sak; 71]Saksonov: On the group ring of finite groups I; Publ. Math. Debrecen 18 (1971), 187–209.

[Sak; 66]Saksonov A.I.: Certain integer–valued rings associated with a finite group; Dokl. Akad. Nauk SSSR 171 (1966), 529–532 = Soviet Math. Dokl. 7 (1966), 1513–1516.

[Sc; 85]Scott L.L.: Letter to K.W. Roggenkamp from June 13, 1985.

[Sc; 87]Scott L.L.: Recent Progress on the isomorphism problem. Proc. of Symposia in Pure Math., Vol. 47 1987, 259–273.

[Sc; 90]Scott L.L.: Réprésentations linéaires des groupes finis, Proc. Colloq. Luminy, France (1988), Astérisque 181–182, (1990).

[Se; 78]Sehgal S.K.: Topics in Group Rings, Marcel Decker, N.Y. 1978.

[Sch; 83]Schmidt P.: Lifting Modular Representations of p–Solvable Groups, J. Alg. 83 (1983), 461–470.

[Sch; 88 1]Schmidt P.: Clifford Theory of Simple Modules, J. Alg. 119 (1988), 185–212.

[Sch; 88 2]Schmidt P.: Extensions of lattices over p–solvable groups, Arch. Math. 50 (1988), 492–494.

[Se; 83]Sehgal S.K.: Torsion units in integral group rings; Proc. Nato Institute on Methods in Ring Theory, Antwerp, D. Riedel, Dordrecht, 1983, pp. 497–504.

[Th; 89]Thompson G.: On the conjugacy of group bases, Ph.D. theses, Univ. of Virginia, (1989).

[We; 87]Weiss A.: p–adic rigidity of p–torsion; Annals of Mathematics, 317–332 (1987).

[Wall; 47]Wall G.E.: Journal of the London Mathematical Society 22, (1947), 315–320.

[Wallace; 87]Wallace D.: On the center and residual finiteness of the automorphism of a group ring. Proc. of the Edinburgh Math Soc. (1987) 30, 207-213.

[Wh; 68]Whitcomb: The group ring problem. Ph. D. thesis, University of
        Chicago (1968).

[Zi; 90]Zimmermann A.: Das Isomorphieproblem ganzzahliger Gruppenringe
        für Gruppen mit abelschem Normalteiler und Quotienten, der eine
        Vermutung von Hans Zassenhaus erfüllt, Diplomarbeit, Universität
        Stuttgart, (1990).

Index

DMV-Seminar Part 2
HOPF ORDERS AND GALOIS MODULE STRUCTURE

M. J. Taylor
with contributions by N. P. Byott

Table of Contents

I   Introduction and Review of the Tame Case

§1   The Basic Problem

The motivation behind the material presented in these notes is the following question:- if N/K is a finite Galois extension of number fields with Galois group Γ, and if $\mathfrak{O}$ and $\mathfrak{O}_N$ are the rings of algebraic integers in K and N respectively, then what can be said about $\mathfrak{O}_N$ as a Γ-module? A complete answer to this would be a description of $\mathfrak{O}_N$ as a module over the group ring $\mathfrak{O}\Gamma$, but since in general $\mathfrak{O}_N$ need not be free over $\mathfrak{O}$, it is more fruitful to restrict scalars and view $\mathfrak{O}_N$ as a $\mathbb{Z}\Gamma$-module.

The following algebraic result is useful in this context:

1.1. Theorem. *(Swan) Let M be a finitely generated $\mathfrak{O}\Gamma$-module with no $\mathbb{Z}$-torsion. Then the following are equivalent:*

   (i) M is $\mathfrak{O}\Gamma$-projective;

   (ii) M is a locally free $\mathfrak{O}\Gamma$-module.

PROOF: See [6, 32.11]   □

For any nonzero prime ideal $\mathfrak{p}$ of $\mathfrak{O}$ we write $e_{\mathfrak{p}}$ for the ramification index of (any prime of N above) $\mathfrak{p}$ in N/K. The extension N/K is said to be *at most tamely ramified*, or, more briefly, *tame*, if for every rational prime p and for every prime ideal $\mathfrak{p}$ of $\mathfrak{O}$ above p, $(e_{\mathfrak{p}}, p) = 1$. Otherwise, N/K is said to be *wild*. The next result gives several characterisations of tame extensions:

1.2. Theorem. *With the above notation, the following are equivalent:*

   (i) N/K is tame;

   (ii) $\mathrm{Tr}_{N/K}(\mathfrak{O}_N) = \mathfrak{O}$;

   (iii) $\mathfrak{O}_N$ is $\mathfrak{O}\Gamma$-projective;

   (iv) $\mathfrak{O}_N$ is $\mathbb{Z}\Gamma$-projective.

In the wild case, we must therefore replace $\mathfrak{O}\Gamma$ by a larger ring of operators if we are to have any hope of proving projectivity results. For tame extensions, on the other hand, we have to describe the global structure of the locally free $\mathbb{Z}\Gamma$-module $\mathfrak{O}_N$. In the remainder of this introductory chapter, we indicate how this structure is governed by the behaviour of certain L-functions associated to N/K. In the subsequent chapters, we will see how the theory of Hopf orders and their principal homogeneous spaces enables us to show that, in certain other situations, the Galois module structure is again dominated by suitable L-functions.

§2   Grothendieck Groups and Classgroups

In order to measure the deviation of a locally free module from being free we introduce a classgroup, analogous to the usual ideal classgroup of a number field. For our discussion of tame extensions N/K we only need the classgroup $Cl(\mathbb{Z}\Gamma)$ of locally free $\mathbb{Z}\Gamma$-modules, but for future use we will for the moment work in greater generality.

Thus let K be a number field, let Γ be a finite group, and let $\mathfrak{A}$ be an order in the group algebra $K\Gamma$:- recall that this means that $\mathfrak{A}$ is a subring of $K\Gamma$, finitely generated as a module over the ring of integers $\mathfrak{O}$ of K, and such that $\mathfrak{A}K = K\Gamma$. In our application to tame extensions, for example, $K = \mathbb{Q}$ and $\mathfrak{A} = \mathbb{Z}\Gamma$.

We define the Grothendieck group $K_0(\mathfrak{A})_{lf}$ of locally free $\mathfrak{A}$-modules to be the abelian group generated by the symbols $[M]$, one for each isomorphism class of locally free $\mathfrak{A}$-modules M, subject to the relations

$$[M \oplus N] - [M] - [N] = 0 \text{ for all classes } [M], [N].$$

By definition, a locally free module M is finitely generated and has a well-defined rank, which is additive with respect to direct sums. Thus there is a rank map

$$\mathrm{rk} : K_0(\mathfrak{A})_{lf} \longrightarrow \mathbb{Z}$$

which is obviously surjective. We define the classgroup $Cl(\mathfrak{A})$ to be its kernel. This is always a finite abelian group, and writing it multiplicatively, we have the short exact sequence

$$1 \longrightarrow Cl(\mathfrak{A}) \longrightarrow K_0(\mathfrak{A})_{lf} \longrightarrow \mathbb{Z} \longrightarrow 0.$$

This sequence has a canonical splitting given by $n \mapsto n[\mathfrak{A}]$, and hence any locally free $\mathfrak{A}$-module M determines a class (M) in $Cl(\mathfrak{A})$, given by the formula

$$(M) = [M] - \mathrm{rk}(M)\,[\mathfrak{A}].$$

If M and N are two locally free $\mathfrak{A}$-modules of the same rank, then $(M) = (N)$ if and only if M and N are stably isomorphic, i.e.

$$M \oplus \mathfrak{A}^n = N \oplus \mathfrak{A}^n$$

for some integer n (where $\mathfrak{A}^n$ denotes the direct sum of n copies of $\mathfrak{A}$). In general this does not imply that M and N are isomorphic, although one does have such an implication if $K\Gamma$ satisfies the Eichler condition (cf. [18, (34.3), (38.1)]).

The above definition of $Cl(\mathfrak{A})$ is not well-suited to calculations, and to obtain concrete results about the classes of specific modules we need an alternative description. First we fix some notation:- let $\mathfrak{g} = (0)$ be the generic (zero) prime ideal of $\mathfrak{O}$, and for each prime ideal $\mathfrak{p} \neq \mathfrak{g}$ of $\mathfrak{O}$ let $\mathfrak{O}_\mathfrak{p}$ denote the completion of $\mathfrak{O}$ at $\mathfrak{p}$. Let

$$\widetilde{\mathfrak{O}} = \prod_{\mathfrak{p} \neq \mathfrak{g}} \mathfrak{O}_\mathfrak{p}$$

$$\mathbb{A} = \widetilde{\mathfrak{O}} \otimes_\mathfrak{O} K$$

Thus $\widetilde{\mathfrak{O}}$ is the ring of finite integral adeles of K (i.e. excluding the infinite places of K), and $\mathbb{A}$ is the full ring of finite adeles of K. For any $\mathfrak{O}$-module M, we will write $M_\mathfrak{p}$ for $M \otimes_\mathfrak{O} \mathfrak{O}_\mathfrak{p}$ and $\widetilde{M}$ for $M \otimes_\mathfrak{O} \widetilde{\mathfrak{O}}$.

Let R_Γ denote the group of virtual characters of Γ, i.e. the free abelian group on the irreducible complex-valued characters of Γ. Also let F be a finite Galois extension of K such that every complex-valued character of Γ is afforded by a representation over F, and set $\Omega = \Omega(K) = \mathrm{Gal}(F/K)$. We will write J_F for the group of finite ideles of F, so $J_F = (\mathbb{A} \otimes_K F)^\times$.

Now let B be a commutative $\mathfrak{O}$-algebra with no $\mathfrak{O}$-torsion:- for instance we may take B to be $\widetilde{\mathfrak{O}}$, K or $\mathbb{A}$. Given $b \in B\Gamma^\times$, we will define a group homomorphism

$$\mathrm{Det}(b) : R_\Gamma \longrightarrow (B \otimes_\mathfrak{O} F)^\times.$$

These determinant maps will allow us to reformulate the definition of $\mathrm{Cl}(\mathfrak{A})$.

First, let $\chi \in R_\Gamma$ be an actual character, afforded by the representation $T : \Gamma \longrightarrow GL_n(F)$. Then T extends to a group homomorphism $T' : B\Gamma^\times \longrightarrow GL_n(B \otimes F)$, and we define $\mathrm{Det}(b)(\chi)$ to be the determinant $\det(T'(b)) \in (B \otimes F)^\times$. This is independent of the choice of representation T affording χ. Next, for any actual characters χ and ϕ, set $\mathrm{Det}(b)(\chi - \phi) = \mathrm{Det}(b)(\chi)\,\mathrm{Det}(b)(\phi)^{-1}$. This gives a well-defined map $\mathrm{Det}(b)$ on the whole of R_Γ, which is easily seen to be a group homomorphism. We have therefore constructed a map

$$\mathrm{Det} : B\Gamma^\times \longrightarrow \mathrm{Hom}(R_\Gamma, (B \otimes F)^\times).$$

We regard R_Γ and $(B \otimes F)^\times$ as Ω-modules, by letting Ω act on R_Γ via the character values and on $B \otimes F$ via the second tensor factor. Writing $\mathrm{Hom}_\Omega(R_\Gamma, (B \otimes F)^\times)$ for the Ω-equivariant homomorphisms $R_\Gamma \longrightarrow (B \otimes F)^\times$, we then have

2.1. Lemma. Det *maps* $B\Gamma^\times$ *into* $\mathrm{Hom}_\Omega(R_\Gamma, (B \otimes F)^\times)$. *In particular,* $\mathrm{Det}(\mathbb{A}\Gamma^\times) \subseteq \mathrm{Hom}_\Omega(R_\Gamma, J_F)$.

PROOF: We must show that $\mathrm{Det}(b)(\chi^\omega) = (\mathrm{Det}(b)(\chi))^\omega$ for all $b \in B\Gamma^\times$, $\chi \in R_\Gamma$, $\omega \in \Omega$. It is sufficient to verify this for actual characters χ. Suppose that χ is afforded by the representation T. Then χ^ω is afforded by T^ω, and writing $b = \sum b_\gamma \gamma$ we have

$$\mathrm{Det}(b)(\chi^\omega) = \det\left(\sum b_\gamma\, T^\omega(\gamma)\right)$$

$$
\begin{aligned}
&= \ \det\left(\sum b_\gamma\, T(\gamma)\right)^\omega \\
&= \ \mathrm{Det}(b)(\chi)^\omega.
\end{aligned}
$$

$\square$

Our aim is now to give Fröhlich's Hom-description of the classgroup $\mathrm{Cl}(\mathfrak{A})$ as a certain quotient of the group $\mathrm{Hom}_\Omega(R_\Gamma, J_F)$. We cannot give all the details here:- for a fuller account see [26, §1] or [8, I§2, II§1], although the formulation given in these books is slightly different from that given here, using adele rings with components at the infinite places of K.

Let M be a locally free $\mathfrak{A}$-module. For simplicity we assume that its rank is 1. For each prime ideal $\mathfrak{p}$ of $\mathfrak{O}$ choose $m_\mathfrak{p} \in M_\mathfrak{p}$ such that $M_\mathfrak{p} = m_\mathfrak{p}\mathfrak{A}_\mathfrak{p}$. In particular we have $MK = m_\mathfrak{g}\,K\Gamma$. For each $\mathfrak{p} \neq \mathfrak{g}$ there is a unique $x_\mathfrak{p} \in K_\mathfrak{p}\Gamma^\times$ with $m_\mathfrak{p} = m_\mathfrak{g}\,x_\mathfrak{p}$. Then $x_\mathfrak{p} \in \mathfrak{O}_\mathfrak{p}\Gamma$ for almost all $\mathfrak{p}$, so we may regard $\underline{x} = (x_\mathfrak{p})_{\mathfrak{p}\neq\mathfrak{g}}$ as an element of $A\Gamma^\times$. Thus $\underline{x}$ determines an element $\mathrm{Det}(\underline{x})$ of $\mathrm{Hom}_\Omega(R_\Gamma, J_F)$. This element depends on the choice of bases $m_\mathfrak{p}$:- $m_\mathfrak{g}$ is only determined up to an element of $K\Gamma^\times$ and $(m_\mathfrak{p})_{\mathfrak{p}\neq\mathfrak{g}}$ is only determined up to an element of $\prod_{\mathfrak{p}\neq\mathfrak{g}}\mathfrak{A}_\mathfrak{p}{}^\times = \widetilde{\mathfrak{A}}^\times$. Thus M gives rise to a well-defined element $\pi'(M)$ of the quotient

$$
\frac{\mathrm{Det}(A\Gamma^\times)}{\mathrm{Det}(\widetilde{\mathfrak{A}}^\times)\,\mathrm{Det}(K\Gamma^\times)}.
$$

This construction can be extended to a locally free $\mathfrak{A}$-module of arbitrary rank, and $\pi'(M)$ only depends on the class $(M) \in \mathrm{Cl}(\mathfrak{A})$. In fact π' induces an isomorphism

$$
\pi : \mathrm{Cl}(\mathfrak{A}) \longrightarrow \frac{\mathrm{Det}(A\Gamma^\times)}{\mathrm{Det}(\widetilde{\mathfrak{A}}^\times)\,\mathrm{Det}(K\Gamma^\times)}.
$$

We can rewrite two of the groups appearing in this quotient. If v is a real place of K then the component of $K_v\Gamma = \mathbb{R}\Gamma$ corresponding to an irreducible symplectic character χ is a matrix algebra over the quaternions, and for $b \in K_v\Gamma^\times$, $\mathrm{Det}(b)(\chi)$ is just the reduced norm of b in this component, which is a sum of squares and hence is positive. Thus $\mathrm{Det}(K\Gamma^\times) \subseteq \mathrm{Hom}_\Omega^+(R_\Gamma, F^\times)$, where the latter group consists of those homomorphisms whose values at any irreducible symplectic character χ are real and positive at each place of F above a real place of K. In fact we have $\mathrm{Det}(K\Gamma^\times) = \mathrm{Hom}_\Omega^+(R_\Gamma, F^\times)$ by the Hasse-Schilling norm theorem [18, (33.15)]. More easily, we also have $\mathrm{Det}(A\Gamma^\times) = \mathrm{Hom}_\Omega(R_\Gamma, J_F)$. Putting all this together, we obtain the Hom-description of the classgroup:

2.2. Theorem. *There is an isomorphism of finite groups*

$$
\mathrm{Cl}(\mathfrak{A}) \cong \frac{\mathrm{Hom}_\Omega(R_\Gamma, J_F)}{\mathrm{Det}(\widetilde{\mathfrak{A}}^\times)\,\mathrm{Hom}_\Omega^+(R_\Gamma, F^\times)}.
$$

§3   The Class of $\mathfrak{O}_N$

We now return to the arithmetic question of describing the ring of integers $\mathfrak{O}_N$ of a tame Galois extension N/K of number fields with Galois group Γ. By Theorem 1.2, $\mathfrak{O}_N$ is a locally free $\mathbb{Z}\Gamma$-module. We shall describe its class in $\mathrm{Cl}(\mathbb{Z}\Gamma)$.

To each complex-valued character χ of Γ is associated an extended Artin L-function $\Lambda(s, \chi)$ (see [12]). This is a meromorphic function of the complex variable s, defined in the half-plane $\mathrm{Re}(s) > 1$ as a product of Euler factors over the places of K (including those ramified in N, and the infinite places). It satisfies the functional equation

$$\Lambda(s, \chi) = W(\chi) \, \Lambda(1 - s, \overline{\chi})$$

where $\overline{\chi}$ is the contragredient (complex-conjugate) character to χ, and $W(\chi)$ is a complex constant of absolute value 1, called the *Artin root number* of χ.

It is immediate from the functional equations of $\Lambda(s, \chi)$ and $\Lambda(s, \overline{\chi})$ that $W(\chi) W(\overline{\chi}) = 1$. In particular, if χ is either orthogonal or symplectic, so that $\overline{\chi} = \chi$, then $W(\chi) = \pm 1$. In fact $W(\chi) = +1$ for any orthogonal character χ; this was first conjectured by Serre, and subsequently proved by Fröhlich and Queyrut. For symplectic characters, on the other hand, the value -1 can occur, as was shown by Armitage.

In this tame situation, it can be shown that $W(\chi^\omega) = W(\chi)$ for all symplectic characters χ and for all $\omega \in \Omega$. Thus we may define $t \in \mathrm{Hom}_\Omega(R_\Gamma, J_F)$ by setting

$$t(\chi) = \begin{cases} W(\chi) & \text{if } \chi \text{ is irreducible and symplectic} \\ 1 & \text{if } \chi \text{ is irreducible but not symplectic.} \end{cases}$$

By Theorem 2.2, t determines a class in $\mathrm{Cl}(\mathbb{Z}\Gamma)$, the *Cassou-Noguès root number class*, which we will again denote by t. It is essentially an analytic invariant of N/K, being defined in terms of the functional equations of the extended Artin L-functions, and it is therefore surprising that it should have any connection with the algebraic structure of $\mathfrak{O}_N$. Yet it determines the $\mathbb{Z}\Gamma$-module structure up to stable isomorphism:- the following striking result was originally conjectured by Fröhlich:

3.1. Theorem. *[25] We have the equality of elements of $\mathrm{Cl}(\mathbb{Z}\Gamma)$:*

$$(\mathfrak{O}_N) = t.$$

The proof of this is beyond the scope of these notes, and can be found in [25] or [8]. The second of these references also gives an account of the historical background to the theorem.

We will end this chapter with two consequences of Theorem 3.1:

3.2. Corollary. *If Γ has no irreducible symplectic characters (e.g. if Γ is abelian or of odd order) then $\mathfrak{O}_N$ is $\mathbb{Z}\Gamma$-free.*

PROOF: The root number class is trivial in this case, so Theorem 3.1 shows that $\mathfrak{O}_N$ is stably free. Moreover the Eichler condition holds for group algebras with no irreducible symplectic characters, so in this case every stably free module is free. □

3.3. Corollary. $(\mathfrak{O}_N)$ *is a self-dual class in* $\mathrm{Cl}(\mathbb{Z}\Gamma)$, *i.e.* $(\mathfrak{O}_N) = (\widehat{\mathfrak{O}}_N)$.

PROOF: For any locally free $\mathbb{Z}\Gamma$-module M, its dual is by definition the locally free $\mathbb{Z}\Gamma$-module $\widehat{M} = \mathrm{Hom}_{\mathbb{Z}}(M, \mathbb{Z})$ with Γ-action $a^\gamma(m) = a(m^{\gamma^{-1}})$ for $a \in \widehat{M}$, $m \in M$, $\gamma \in \Gamma$.

If $f \in \mathrm{Hom}_{\Omega(\mathbb{Q})}(R_\Gamma, J_F)$ represents the class (M) then I claim that $(\widehat{M})$ is represented by $\widehat{f}$, where $\widehat{f}(\chi) = f(\overline{\chi})^{-1}$. As $\widehat{t} = t$ for the root number class t, this will prove the corollary.

For simplicity, we prove the claim only for M of rank 1, and assume without loss of generality that M is a locally free $\mathbb{Z}\Gamma$-ideal. Thus we can find an idele $\underline{\alpha} = (\alpha_p)$ of $\mathbb{Q}\Gamma$ such that $M_p = \alpha_p \mathbb{Z}_p\Gamma$ for all p. Now define a $\mathbb{Q}$-bilinear pairing

$$< \cdot, \cdot > : \mathbb{Q}\Gamma \times \mathbb{Q}\Gamma \longrightarrow \mathbb{Q}$$

by

$$< \gamma, \delta > = \begin{cases} 1 & \text{if } \gamma = \delta \\ 0 & \text{otherwise} \end{cases}$$

for all γ, $\delta \in \Gamma$. For each prime number p, this extends to a $\mathbb{Q}_p$-bilinear pairing on taking p-adic completions, and so gives rise to a surjective $\mathbb{Z}_p$-bilinear pairing

$$< \cdot, \cdot >_p : \alpha_p \mathbb{Z}_p\Gamma \times \overline{\alpha}_p^{-1} \mathbb{Z}_p\Gamma \longrightarrow \mathbb{Z}_p$$

where, for any $\alpha = \sum \alpha_\gamma \gamma \in \mathbb{Q}_p\Gamma$ we write $\overline{\alpha}$ for $\sum \alpha_\gamma \gamma^{-1}$. This identifies $\widehat{M}_p$ with $\overline{\alpha}_p^{-1} \mathbb{Z}_p\Gamma$ as a $\mathbb{Z}_p\Gamma$-module. Thus by the discussion preceding Theorem 3.1, we may take as representatives of (M) and $(\widehat{M})$ the maps $f = \mathrm{Det}(\underline{\alpha})$ and $\widehat{f} = \mathrm{Det}(\overline{\underline{\alpha}}^{-1})$ in $\mathrm{Hom}_\Omega(R\Gamma, J_F)$. Now

$$\begin{aligned} \widehat{f}(\chi) &= \mathrm{Det}(\overline{\underline{\alpha}}^{-1})(\chi) \\ &= (\mathrm{Det}(\overline{\underline{\alpha}})(\chi))^{-1} \\ &= (\mathrm{Det}(\underline{\alpha})(\overline{\chi}))^{-1} \\ &= f(\overline{\chi})^{-1} \end{aligned}$$

as claimed. □

II Hopf Orders

To study the integral Galois module structure of wildly ramified extensions of number fields, we will need to work with orders in the group algebra that are larger than the group ring. We shall consider only those orders of a special type, namely Hopf orders. We will only be concerned with orders in either a group algebra or the algebra of maps from a group to the ground field, and for ease of exposition we will only define Hopf orders in the context of these particular Hopf algebras. After giving various general algebraic results about such Hopf algebras and Hopf orders, we will show how to classify all Hopf orders in the group algebra of a group of prime order, and will briefly discuss how this technique can be generalised to a special class of Hopf orders in the group algebra of an elementary abelian group. In the final section of this chapter, we will indicate how commutative Hopf algebras in general have a geometric interpretation in terms of affine group schemes, and will show how examples of Hopf orders can be constructed using formal groups.

§1 Hopf Orders in KG

Until further notice, $\mathfrak{O}$ denotes a Dedekind domain, K its field of fractions, which is assumed to have characteristic 0, and G a finite abelian group of order n. We write A for the group algebra KG.

We define K-linear maps

$$\begin{aligned} \Delta &: A \longrightarrow A \otimes_K A \quad \text{(comultiplication)} \\ \epsilon &: A \longrightarrow K \quad \text{(augmentation)} \\ {}^{-} &: A \longrightarrow A \quad \text{(involution)} \end{aligned}$$

on group elements $g \in G$ by

$$\Delta(g) = g \otimes g, \quad \epsilon(g) = 1, \quad \bar{g} = g^{-1}.$$

These are all algebra homomorphisms.

If $\mathfrak{A}$ is an order in A then we view $\mathfrak{A} \otimes_{\mathfrak{O}} \mathfrak{A}$ as an order in $A \otimes_K A$. We call the order $\mathfrak{A}$ a *Hopf order* if $\Delta(\mathfrak{A}) \subseteq \mathfrak{A} \otimes_{\mathfrak{O}} \mathfrak{A}$.

If this condition holds, we say that $\mathfrak{A}$ is costable (under Δ), by analogy with the fact that a lattice $\mathfrak{L}$ in A is said to be stable (under multiplication) if $\mathfrak{L}\mathfrak{L} \subseteq \mathfrak{L}$.

For any $m \in \mathbb{Z}$, let $[m] : A \longrightarrow A$ be the K-linear map given by $[m](g) = g^m$.

1.1. Lemma. *For any Hopf order $\mathfrak{A}$ in A and any $m \in \mathbb{Z}$, we have $[m](\mathfrak{A}) \subseteq \mathfrak{A}$.*

PROOF: Adding a suitable multiple of n to m if necessary, we may assume

$m > 0$. Consider the composition

$$A \xrightarrow{\Delta} A \otimes A \xrightarrow{1 \otimes \Delta} A \otimes A \otimes A \xrightarrow{1 \otimes 1 \otimes \Delta} \cdots \longrightarrow \underbrace{A \otimes \cdots \otimes A}_{m} \xrightarrow{\text{multiply}} A.$$

This takes $\mathfrak{A}$ into $\mathfrak{A}$, since $\mathfrak{A}$ is both costable under Δ and stable under multiplication. But this map is K-linear, and takes a group element g to g^m, so it coincides with $[m]$. $\qquad\qquad\square$

1.2. Corollary. *If $\mathfrak{A}$ is a Hopf order in A then $\overline{\mathfrak{A}} = \mathfrak{A}$ and $\epsilon(\mathfrak{A}) = \mathfrak{O}$.*

PROOF: For $a \in \mathfrak{A}$, $\bar{a} = [n-1](a) \in \mathfrak{A}$. Thus $\overline{\mathfrak{A}} \subseteq \mathfrak{A}$, and as $\bar{\bar{a}} = a$, the inclusion must be equality. Also $\epsilon(a) = [n](a) \in \mathfrak{A} \cap K = \mathfrak{O}$, the last equality holding as $\mathfrak{A}$ is an order in A. Thus $\epsilon(\mathfrak{A}) \subseteq \mathfrak{O}$, and as $\epsilon(\mathfrak{A}) \supseteq \epsilon(\mathfrak{O}) = \mathfrak{O}$ we again have equality. $\qquad\qquad\square$

1.3. Examples.

(i): The integral group ring $\mathfrak{O}G$ is a Hopf order:- clearly $\Delta(\mathfrak{O}G) \subseteq \mathfrak{O}G \otimes_{\mathfrak{O}} \mathfrak{O}G$.

(ii): Let $\mathfrak{M}$ be the unique maximal order in KG. If the discriminant $\mathfrak{d}(\mathfrak{M})$ of $\mathfrak{M}$ (with respect to the trace pairing) is $\mathfrak{O}$ then $\mathfrak{M}$ is a Hopf order (and hence the maximal Hopf order).

PROOF OF (ii): Consider the order $\mathfrak{M} \otimes_{\mathfrak{O}} \mathfrak{M}$ in the algebra $A \otimes_K A$:- an easy calculation shows that $\mathfrak{d}(\mathfrak{M} \otimes \mathfrak{M}) = \mathfrak{d}(\mathfrak{M})^{2n} = \mathfrak{O}$, so it is the (unique) maximal order. Since Δ is an algebra homomorphism, it follows that $\Delta(\mathfrak{A}) \subseteq \mathfrak{M} \otimes \mathfrak{M}$ for any order $\mathfrak{A}$ in A. Taking $\mathfrak{A} = \mathfrak{M}$, this shows that $\mathfrak{M}$ is a Hopf order. $\qquad\square$

ALTERNATIVE PROOF OF (ii): As G is abelian, the algebra $A = KG$ is a product of fields. Hence $\mathfrak{M}$ is the product of the integral closures of $\mathfrak{O}$ in these fields, and $\mathfrak{d}(\mathfrak{M})$ is the product of the discriminants of these field extensions of K, viewed as an $\mathfrak{O}$-ideal. Thus if $\mathfrak{d}(\mathfrak{M}) = \mathfrak{O}$, each of these extensions is unramified at every prime of $\mathfrak{O}$. Replacing A by $A \otimes_K K'$ for a suitable unramified extension K' of K, so that $\mathfrak{M}$ is replaced by $\mathfrak{M} \otimes_{\mathfrak{O}} \mathfrak{O}'$ (where $\mathfrak{O}'$ is the integral closure of $\mathfrak{O}$ in K'), we may therefore assume that

$$A = \prod_{\chi} K,$$

where the product runs over all abelian characters χ of G. Explicitly,

$$A = \prod_{\chi} K \, e_{\chi}$$

where the idempotents e_{χ} are given by

$$e_{\chi} = \frac{1}{n} \sum_{g \in G} \chi(g^{-1}) \, g.$$

Then clearly $\mathfrak{M} = \prod_\chi \mathfrak{O}\, e_\chi$. Now

$$\Delta(e_\chi) = \sum_{\phi\psi=\chi} e_\phi \otimes e_\psi$$

as is easily checked using the orthogonality relations for abelian characters. Thus $\Delta(\mathfrak{M}) \subseteq \mathfrak{M} \otimes \mathfrak{M}$, as required. □

§2   The Childs-Hurley Criterion

Hopf orders in the group algebra $A = KG$ have a number of properties in common with the integral group ring $\mathfrak{O}G$. As a first example of this, we will prove a projectivity criterion for modules over a Hopf order, generalising a well-known result for group rings (cf. [5, (62.3)]).

We will write Σ for the element $\sum_{g \in G} g$ of A, and $K\Sigma$ for the 1-dimensional subspace of A that it generates.

2.1. Theorem. (*Childs-Hurley [3]*) *Suppose that $\mathfrak{O}$ is a principal ideal domain and let $\mathfrak{A}$ be a Hopf order in A. Let c be a generator of the $\mathfrak{O}$-ideal $\mathfrak{c}$ defined by $\mathfrak{A} \cap K\Sigma = \mathfrak{c}^{-1}\Sigma$. Let M be a finitely generated $\mathfrak{A}$-module with no $\mathfrak{O}$-torsion. If there exists an element f of $\operatorname{End}_{\mathfrak{O}}(M)$ such that*

$$\sum_{g \in G} f(m\,g)\, g^{-1} = c\,m \ \text{ for all } m \in M$$

then M is $\mathfrak{A}$-projective.

PROOF: Let $d = c^{-1}\,\Sigma$. Then $d \in \mathfrak{A}$, and as $\mathfrak{A}$ is a Hopf order we have

$$\Delta(d) = \sum_{i=1}^{r} d_i^{(1)} \otimes d_i^{(2)} \ \text{ for some } d_i^{(1)},\, d_i^{(2)} \in \mathfrak{A}.$$

For $m \in M$ set

$$\lambda(m) = \sum_{i=1}^{r} f(m\, d_i^{(1)}) \otimes \overline{d_i^{(2)}}.$$

Since M admits $\mathfrak{A}$ and since $\overline{\mathfrak{A}} = \mathfrak{A}$, this defines an $\mathfrak{O}$-linear map $\lambda : M \longrightarrow M \otimes_{\mathfrak{O}} \mathfrak{A}$, which is independent of the choice of expansion of $\Delta(d)$.

Working in A, we can write

$$\Delta(d) = c^{-1} \sum_{g \in G} g \otimes g.$$

Thus

$$\lambda(m) = c^{-1} \sum_{g \in G} f(mg) \otimes g^{-1}.$$

The map $\mu : M \otimes_{\mathfrak{D}} \mathfrak{A} \longrightarrow M$ given by $\mu(m \otimes a) = m\,a$ splits λ since

$$\mu \circ \lambda(m) = c^{-1} \sum_{g \in G} f(m\,g)\,g^{-1} = m$$

by the hypothesis on f. It is easily verified that λ and μ are G-equivariant, and are therefore $\mathfrak{A}$-module maps, where $M \otimes_{\mathfrak{D}} \mathfrak{A}$ is viewed as an $\mathfrak{A}$-module by multiplication on the right. We have thus shown that M is a direct summand of the free $\mathfrak{A}$-module $M \otimes_{\mathfrak{D}} \mathfrak{A}$. It is therefore $\mathfrak{A}$-projective. $\square$

We will need to know when a projective $\mathfrak{A}$-module M is locally free over $\mathfrak{A}$. Clearly this can only be the case if M spans a free A-module. Conversely, we have the following result:

2.2. Lemma. *Let $\mathfrak{A}$ be any order in A, and let M be a finitely generated $\mathfrak{A}$-module with no $\mathfrak{D}$-torsion, which spans a free A-module. Then M is locally free over $\mathfrak{A}$ if and only if M is projective over $\mathfrak{A}$.*

PROOF: Since G is abelian and K has characteristic 0, A is a commutative separable K-algebra. The result then follows from the equivalence (iv) $\leftrightarrow$ (v) of [7, Theorem 4]. $\square$

Using Theorem 2.1, we will now obtain an analogue of Theorem 1.2 of Chapter I for a wild abelian extension whose ring of integers admits a Hopf order:

2.3. Theorem. *Let N/K be a finite abelian extension of number fields, with Galois group Γ, and let $\mathfrak{D}_N$ and $\mathfrak{D} = \mathfrak{D}_K$ be the rings of integers of N and K. Suppose that $\mathfrak{D}_N$ admits a Hopf order $\mathfrak{A}$ in $K\Gamma$. Writing*

$$\Sigma = \sum_{g \in G} g \in K\Gamma,$$

let $\mathfrak{c}$ be the $\mathfrak{D}$-ideal such that

$$\mathfrak{c}\,(\mathfrak{A} \cap K\Sigma) = \mathfrak{D}\Sigma.$$

Then $\mathfrak{D}_N$ is $\mathfrak{A}$-projective if and only if $\mathrm{Tr}_{N/K}(\mathfrak{D}_N) = \mathfrak{c}$.

PROOF: The two conditions to be shown equivalent are preserved under localisation, so we may assume that $\mathfrak{D}$ is a principal ideal domain. Let $\mathfrak{c} = c\mathfrak{D}$. Then the element c is as in the statement of Theorem 2.1.

If $\mathrm{Tr}_{N/K}(\mathfrak{D}_N) = \mathfrak{c}$ then there exists $x \in \mathfrak{D}_N$ with $\mathrm{Tr}_{N/K}(x) = c$. Pick such an x and define $f(m) = x\,m$ for $m \in \mathfrak{D}_N$. Then

$$\sum_{g \in G} f(mg^{-1})g \;=\; \sum_{g \in G} (x\,(mg^{-1}))g$$

$$=\; \sum_{g \in G} x^g\,m$$

$$= m \operatorname{Tr}_{L/K}(x)$$
$$= cm$$

and $\mathfrak{O}_N$ is $\mathfrak{A}$-projective by Theorem 2.1.

Conversely, suppose that $\mathfrak{O}_N$ is $\mathfrak{A}$-projective. Since $\mathfrak{O}_N$ spans the free $K\Gamma$-module N, it is locally free over $\mathfrak{A}$ by Lemma 2.2. Thus an element α of $K\Gamma$ lies in $\mathfrak{A}$ if and only if $\mathfrak{O}_N \alpha \subseteq \mathfrak{O}_N$. For any nonzero ideal $\mathfrak{r}$ of $\mathfrak{O}$ we therefore have

$$\mathfrak{O}\Sigma \subseteq \mathfrak{r}(\mathfrak{A} \cap K\Sigma) \quad \Leftrightarrow \quad \mathfrak{r}^{-1}\Sigma \subseteq \mathfrak{A}$$
$$\Leftrightarrow \quad \mathfrak{r}^{-1}\mathfrak{O}_N\Sigma \subseteq \mathfrak{O}$$
$$\Leftrightarrow \quad \operatorname{Tr}_{N/K}(\mathfrak{O}_N) \subseteq \mathfrak{r},$$

so $\operatorname{Tr}_{N/K}(\mathfrak{O}_N) = \mathfrak{c}$. $\qquad\qquad\qquad\qquad\qquad\qquad\qquad\qquad\qquad\qquad$ □

§3 Duality at Field Level

Let $B = \operatorname{Map}(G, K)$ be the K-vector space of maps from G to K. We view this as a K-algebra with pointwise addition and multiplication of maps:- explicitly, for f, $f' \in B$ and $g \in G$ we set

$$(f + f')(g) = f(g) + f'(g), \quad (f\,f')(g) = f(g)\,f'(g)$$

Identifying $B \otimes_K B$ with the algebra $\operatorname{Map}(G \times G, K)$ in the obvious manner, we define a comultiplication map $\Delta_B : B \longrightarrow B \otimes B$ by

$$\Delta_B(f)(g, h) = f(gh) \text{ for } f \in B, \ g, h \in G.$$

We also define maps $\epsilon : B \longrightarrow K$ and $^- : B \longrightarrow B$ by

$$\epsilon(f) = f(1_G), \quad \overline{f}(g) = f(g^{-1}) \text{ for } f \in B, \ g \in G.$$

There is a natural non-singular K-bilinear pairing

$$\langle \,.\, , \,.\, \rangle : B \times A \longrightarrow K$$

given by

$$\left\langle f, \sum_g a_g\, g \right\rangle = \sum_g a_g\, f(g)$$

and we can therefore view B as the dual vector space $A^D = \operatorname{Hom}_K(A, K)$ of A.

Recall that if V and W are finite-dimensional vector spaces over K and if $\alpha : V \longrightarrow W$ is any K-linear map then α induces a K-linear map $\alpha^D : W^D \longrightarrow V^D$, given by $\alpha^D(x) = x \circ \alpha$ for $x \in W^D$. Moreover $(V^D)^D$ identifies canonically with V, and then $(\alpha^D)^D = \alpha$. Also $(V \otimes_K W)^D$ identifies canonically with $V^D \otimes_K W^D$.

The multiplication and comultiplication maps $m_A : A \otimes A \longrightarrow A$ and $\Delta_A : A \longrightarrow A \otimes A$ thus give rise to maps $m_A{}^D : B \longrightarrow B \otimes B$ and $\Delta_A{}^D : B \otimes B \longrightarrow B$.

3.1. Proposition.
$$m_A{}^D = \Delta_B \quad and \quad \Delta_A{}^D = m_B.$$

PROOF: For f, $f' \in B$ and g, $g' \in G$ we have

$$m_A{}^D(f)(g,g') = (f \circ m_A)(g,g') = f(gg') = \Delta_B(f)(g,g')$$

and

$$\Delta_A{}^D(f \otimes f')(g) = ((f \otimes f') \circ \Delta_A)(g) = (f \otimes f')(g \otimes g) = f(g)\,f'(g) = (ff')(g).$$

$\square$

Applying the functor $(.)^D$ to Proposition 3.1, we likewise have

3.2. Proposition.
$$m_B{}^D = \Delta_A \quad and \quad \Delta_B{}^D = m_A.$$

To summarise, duality interchanges multiplication and comultiplication in the Hopf algebras A and B.

We view B as an A-module by the rule $f^g(h) = f(h\,g^{-1})$ for $f \in B$, g, $h \in G$. Thus for any a, $a' \in A$ we have $f^a(a') = f(a'\,\bar{a})$. The action of A on B can be written in terms of the duality pairing as

$$< f^g, hg > \, = \, < f, h > \quad \text{for } f \in B,\ g,h \in G.$$

Thus the pairing $< ., . >$ is a G-invariant form (where G acts on A by right multiplication).

Let ℓ denote the element of B given by

$$\ell(h) = \begin{cases} 1 & \text{if } h = 1_G \\ 0 & \text{otherwise.} \end{cases}$$

Then, for any g, $h \in G$,

$$\ell^g(h) = \begin{cases} 1 & \text{if } h = g \\ 0 & \text{otherwise.} \end{cases}$$

Thus the elements $(\ell^g)_{g \in G}$ form a K-basis of B, dual to the basis of A consisiting of the group elements. The ℓ^g are a full set of primitive idempotents in the algebra B, so

$$B = \prod_{g \in G} K\,\ell^g;$$

explicitly, for any $f \in B$ we have

$$f = \sum_{g \in G} f(g)\, \ell^g.$$

§4   Duality at Integral Level

If $\mathfrak{B}$ is an order in B then, as before, we say that it is a Hopf order if $\Delta(\mathfrak{B}) \subseteq \mathfrak{B} \otimes_{\mathfrak{O}} \mathfrak{B}$. The maps $g \mapsto g^m$ for positive integers m induce K-linear endomorphisms of B, and again it is easily verified that any Hopf order admits these maps. We will now show how the discussion of the previous section carries over to integral level, giving duality between Hopf orders in A and B.

Let $\mathfrak{A}$ be a Hopf order in A. Its $\mathfrak{O}$-linear dual $\mathfrak{A}^D = \mathrm{Hom}_{\mathfrak{O}}(\mathfrak{A}, \mathfrak{O})$ identifies in a natural way with a lattice $\mathfrak{B}$ in B. We now verify that $\mathfrak{B}$ is a Hopf order in B.

Firstly, if $a \in A$ then

$$1_B(a) \;=\; \langle \sum_g \ell^g \,,\, a \rangle \;=\; \epsilon(a) \in \mathfrak{O},$$

so $1 \in \mathfrak{B}$. Next, if $b, b' \in \mathfrak{B}$ and $a \in \mathfrak{A}$ then

$$(bb')(a) = (m_B(b \otimes b'))(a) = ((b \otimes b') \circ m_B{}^D)(a) = (b \otimes b')(\Delta(a))$$

by Proposition 3.2, and since $\Delta(a) \in \mathfrak{A} \otimes \mathfrak{A}$, this is in $\mathfrak{O}$. Thus $\mathfrak{B}$ is stable under multiplication, and is therefore an order in B. Finally, if $b \in \mathfrak{B}$ and $a, a' \in \mathfrak{A}$ then

$$(\Delta(b))(a \otimes a') = (m_A{}^D(b))(a \otimes a') = (b \circ m)(a \otimes a') = b(aa') \in \mathfrak{O}$$

since $\mathfrak{A}$ is stable under multiplication. Thus $\Delta(\mathfrak{B}) \subseteq (\mathfrak{A} \otimes_{\mathfrak{O}} \mathfrak{A})^D$, which identifies with the lattice $\mathfrak{B} \otimes_{\mathfrak{O}} \mathfrak{B}$ in $B \otimes_K B$. Hence $\mathfrak{B}$ is indeed a Hopf order in B.

In brief, the above argument says that, since the Hopf order $\mathfrak{A}$ is preserved under both multiplication and comultiplication, and these operations are interchanged under duality, the dual of $\mathfrak{A}$ is preserved under these operations and so is again a Hopf order.

Similarly, given a Hopf order $\mathfrak{B}$ in B, its dual is a Hopf order $\mathfrak{A}$ in A. Clearly $(\mathfrak{A}^D)^D = \mathfrak{A}$ and $(\mathfrak{B}^D)^D = \mathfrak{B}$.

4.1. Example. Take $\mathfrak{A} = \mathfrak{O}G$. Then for $b \in B$,

$$< b, \mathfrak{A} > \,\in \mathfrak{O} \Leftrightarrow b(g) \in \mathfrak{O} \ \text{ for all } g \in G,$$

so $(\mathfrak{O}G)^D = \mathrm{Map}(G, \mathfrak{O})$, which is therefore a Hopf order in B. In terms of the basis of idempotents $(\ell^g)_{g \in G}$ we have

$$\mathrm{Map}(G, \mathfrak{O}) = \prod_{g \in G} \mathfrak{O}\, \ell^g$$

which is visibly the unique maximal order in the algebra

$$B = \prod_{g \in G} K \, \ell^g,$$

and hence is the unique maximal Hopf order in B.   □

Since inclusions of lattices are reversed under duality, we have as an immediate consequence

4.2. Proposition. $\mathfrak{O}G$ *is the unique minimal Hopf order in* $A = KG$.

If $\mathfrak{A}$ is a Hopf order in A and if $\mathfrak{B} = \mathfrak{A}^D$ then $\mathfrak{B}$ is an $\mathfrak{A}$-module:- the rule for the action of A on B gives

$$b^a(a') = b(a' \, \overline{a}) \ \text{ for } b \in \mathfrak{B}, \ a, a' \in \mathfrak{A},$$

and since $\mathfrak{A}$ is closed under multiplication and involution, the right-hand side is in $\mathfrak{O}$, showing that $b^a \in \mathfrak{B}$. Our next goal is to describe the structure of $\mathfrak{B}$ as an $\mathfrak{A}$-module. Before doing so, we introduce the ideals of integrals in the Hopf orders $\mathfrak{A}$ and $\mathfrak{B}$.

The element

$$\Sigma = \sum_{g \in G} g$$

of A has the property that $g\Sigma = \Sigma$ for all $g \in G$, and hence $a\,\Sigma = \epsilon(a)\,\Sigma$ for all $a \in A$. It is easily checked that this property determines Σ up to a scalar multiple. Thus for any Hopf order $\mathfrak{A}$ in A,

$$\{\lambda \in \mathfrak{A} : a\,\lambda = \epsilon(a)\,\lambda \ \text{for all } a \in A\} = K\Sigma \cap \mathfrak{A}.$$

Let $I = K\Sigma \cap \mathfrak{A}$. Then I is an ideal in $\mathfrak{A}$, called the *ideal of integrals* in $\mathfrak{A}$. We have already seen in Theorem 2.1 that it plays an important role in the module theory of $\mathfrak{A}$.

Analogously, in B we have

$$
\begin{aligned}
&\{f \in B : bf = \epsilon_B(b)\,f \text{ for all } b \in B\} \\
&= \ \{f \in B : bf = b(1)\,f \ \text{for all } b \in B\} \\
&= \ \{f \in B : \ell^g f = \ell^g(1)\,f \ \text{for all } g \in G\} \\
&= \ \{f \in B : \ell f = f, \ \ell^g(f) = 0 \ \text{if } g \neq 1_G\} \\
&= \ K\,\ell.
\end{aligned}
$$

Thus the ideal of integrals in $\mathfrak{B}$ is

$$J = \{f \in \mathfrak{B} : bf = \epsilon_B(b)\,f \ \text{for all } b \in B\} = \mathfrak{B} \cap K\,\ell.$$

The following result is essentially contained in [11], although not stated there explicitly:

4.3. Theorem. *Let $\mathfrak{A}$ be a Hopf order in A, and let $\mathfrak{B} = \mathfrak{A}^D$. Then $\mathfrak{B}$ is locally free as an $\mathfrak{A}$-module (i.e. $\mathfrak{A}$ is locally a Frobenius algebra - cf. [6, (9.4)]).*

More precisely, writing J for the ideal of integrals of $\mathfrak{B}$, the map

$$\alpha : J \otimes \mathfrak{A} \longrightarrow \mathfrak{B}$$

$$\alpha(j \otimes a) = j^a$$

is an isomorphism of $\mathfrak{A}$-modules, where $\mathfrak{A}$ acts on $J \otimes \mathfrak{A}$ by multiplication in the second factor.

PROOF: Clearly α is a map of $\mathfrak{A}$-modules, so we must find a two-sided inverse $\beta : \mathfrak{B} \longrightarrow J \otimes \mathfrak{A}$ to α. We give the construction of β in several steps.

Step 1: View $\mathfrak{B}$ as embedded in $\mathrm{Hom}_{\mathfrak{D}}(\mathfrak{B}, \mathfrak{B})$ by identifying $b \in \mathfrak{B}$ with multiplication by b. Let Ψ be the composite

$$\mathfrak{B} \hookrightarrow \mathrm{Hom}_{\mathfrak{D}}(\mathfrak{B}, \mathfrak{B}) \cong \mathfrak{B} \otimes \mathrm{Hom}_{\mathfrak{D}}(\mathfrak{B}, \mathfrak{D}) = \mathfrak{B} \otimes \mathfrak{A}.$$

Explicitly,

(1) $$\Psi(b) = \sum_i b_i \otimes a_i \in \mathfrak{B} \otimes \mathfrak{A}$$

where, for all $b' \in B$,

$$b b' = \sum_i b_i < b', a_i > .$$

Since the duality pairing is G-invariant, for $g \in G$ and $b' \in B$ we have

$$
\begin{aligned}
b^g b'^g &= (b b')^g \\
&= \sum_i b_i{}^g < b', a_i > \\
&= \sum_i b_i{}^g < b'^g, a_i{}^g >
\end{aligned}
$$

so $\Psi(b^g) = \Psi(b)^g$, with G acting diagonally on $\mathfrak{B} \otimes \mathfrak{A}$.

Step 2: We note that $J = \{b \in \mathfrak{B} : \Psi(b) = b \otimes 1\}$:- for

$$\Psi(b) = b \otimes 1 \iff b b' = b < b', 1 > = b'(1) b \text{ for all } b' \in B$$
$$\iff b \in J.$$

Step 3: Define $q : \mathfrak{B} \longrightarrow \mathfrak{B}$ by

$$q(b) = \sum_i b_i \bar{a}_i,$$

with the notation of (1). Then I claim that

$$\Psi(q(b)) = q(b) \otimes 1.$$

It then follows from Step 2 that $q(\mathfrak{B}) \subseteq J$.

To establish the claim, we may extend scalars to K, and by linearity it is sufficient to verify that $\Psi(q(\ell^g)) = q(\ell^g) \otimes 1$ for each $g \in G$. But $\ell b' = b'(1)\ell$ for $b' \in B$, so

(2) $$\Psi(\ell) = \ell \otimes 1,$$

and hence $\Psi(\ell^g) = \Psi(\ell)^g = \ell^g \otimes g$. Thus $q(\ell^g) = (\ell^g)^{g^{-1}} = \ell$, and by (2), $\Psi(q(\ell^g)) = \ell \otimes 1 = q(\ell^g) \otimes 1$ as required.

Step 4: Let $\beta = (q \otimes 1) \circ \Psi : \mathfrak{B} \longrightarrow J \otimes \mathfrak{A}$. Then β is the required inverse to α:- to verify this we again extend scalars. Since $\alpha(\ell \otimes g) = \ell^g$ and $\Psi(\ell^g) = \ell^g \otimes g$ we have $\beta \circ \alpha(\ell \otimes g) = (q \otimes 1)(\ell^g \otimes g) = \ell \otimes g$, showing that $\beta \circ \alpha = 1$. On the other hand, $\beta(\ell^g) = q(\ell^g) \otimes g = \ell \otimes g$, so $\alpha \circ \beta(\ell^g) = \ell^g$, whence $\alpha \circ \beta = 1$. This concludes the proof of the Theorem. □

Of course we may also consider $\mathfrak{A}$ as a $\mathfrak{B}$-module, and *mutatis mutandis* the above proof applies in this dual situation, so that $\mathfrak{A} \cong I \otimes \mathfrak{B}$ as a $\mathfrak{B}$-module.

We may write the ideals of integrals I and J in $\mathfrak{A}$ and $\mathfrak{B}$ as

$$I = \mathfrak{a}\frac{\Sigma}{n}, \quad J = \mathfrak{b}\,\ell,$$

where n is the order of G and $\mathfrak{a}$, $\mathfrak{b}$ are fractional $\mathfrak{O}$-ideals. Since the elements Σ/n and ℓ are idempotents in A and B respectively, and since $\mathfrak{A}$ and $\mathfrak{B}$ are orders, it follows that $\mathfrak{a}$ and $\mathfrak{b}$ are in fact integral $\mathfrak{O}$-ideals. Theorem 4.3 and its dual enable us to determine how these two ideals are related:

4.4. Proposition. $\mathfrak{ab} = n\mathfrak{O}$.

PROOF: We have shown that $\mathfrak{B} = J\mathfrak{A}$ and $\mathfrak{A} = I\mathfrak{B}$. The duality pairing $\mathfrak{B} \times \mathfrak{A} \longrightarrow \mathfrak{O}$ is G-invariant, and hence $\mathfrak{A}$-invariant. Dually it is $\mathfrak{B}$-invariant. Thus

$$
\begin{aligned}
\mathfrak{O} &= \ <\mathfrak{B}, \mathfrak{A}> \ &=& \ <J\mathfrak{A}, \mathfrak{A}> \\
&= \ <J, \mathfrak{A}^2> \ &=& \ <J, \mathfrak{A}> \\
&= \ <J, I\mathfrak{B}> \ &=& \ <J\mathfrak{B}, I> \\
&= \ <J, I> \ &=& \ <\mathfrak{b}\,\ell, \mathfrak{a}\frac{\Sigma}{n}> \\
&= \ \mathfrak{ab} <\ell, \frac{\Sigma}{n}> \ &=& \ \frac{1}{n}\mathfrak{ab}
\end{aligned}
$$

and the result follows. □

We can also give a formula for the different $\mathcal{D}(\mathfrak{B})$ of the Hopf order $\mathfrak{B}$. Recall that

$$\mathcal{D}(\mathfrak{B})^{-1} = \{b \in B : \mathrm{Tr}(b\mathfrak{B}) \subseteq \mathfrak{O}\}$$

where $\mathrm{Tr} : B \longrightarrow K$ is the trace map.

4.5. Proposition. $\mathcal{D}(\mathfrak{B}) = \mathfrak{b}\mathfrak{B}$.

PROOF: Consider the composite isomorphism of $\mathfrak{A}$-modules

$$\psi : \mathcal{D}(\mathfrak{B})^{-1} \cong \mathrm{Hom}(\mathfrak{B}, \mathfrak{O}) = \mathfrak{B}^D = \mathfrak{A} :$$

explicitly, for $d \in \mathcal{D}(\mathfrak{B})^{-1}$, $\psi(d)$ is determined by

$$\mathrm{Tr}(d\,b) \; = \; < b, \psi(d) > \quad \text{for all } b \in B.$$

Since $\ell\, b = (b(1), 0, \ldots, 0) \in B = \prod K$, we have $\mathrm{Tr}(\ell\, b) = b(1) = < b, 1 >$, so $\psi(\ell) = 1$. Now

$$\psi(\mathfrak{B}) = \psi(\mathfrak{b}\,\ell\,\mathfrak{A}) = \mathfrak{b}\,\psi(\ell)\,\mathfrak{A} = \mathfrak{b}\,\mathfrak{A},$$

so $\mathcal{D}(\mathfrak{B})^{-1} = \psi^{-1}(\mathfrak{A}) = \mathfrak{b}^{-1}\,\mathfrak{B}$, giving the result. $\square$

§5   Hopf Orders of Rank p

We now take G to be a cyclic group of prime order p, and determine all Hopf orders in the group algebra $A = KG$. The argument is an adaptation of that used by Tate and Oort to classify all group schemes of rank p ([24]), which itself depends on a result of Deligne.

To begin with, we take $\mathfrak{O}$ to be the valuation ring of a finite extension K of the p-adic field $\mathbb{Q}_p$. By Hensel's lemma, each non-zero residue class of $\mathbb{Z}_p \bmod p\mathbb{Z}_p$ contains a unique $(p-1)$th root of 1. The *Teichmüller character*

$$\phi : \mathbb{F}_p{}^\times \to \mathbb{Z}_p{}^\times$$

is the map taking the class $a \bmod p$ to the unique solution of

$$x^{p-1} = 1, \quad x \equiv a \pmod{p\mathbb{Z}_p}.$$

This is obviously a group homomorphism.

Recall from Lemma 1.1 that any Hopf order $\mathfrak{A}$ in KG admits the maps $[m]$ given by $g \mapsto g^m$ for $g \in G$. Let $\Lambda = \mathrm{Aut}(G)$ be the group of automorphisms of G, and extend each such automorphism to KG by linearity. Then, identifying $m \bmod p$ with $[m]$, we have

$$\Lambda = \{[m] : 1 \le m \le p-1\} = \mathbb{F}_p{}^\times.$$

The ring $\mathfrak{O}\Lambda$ has a basis of primitive idempotents e_i for $1 \le i \le p-1$, where

$$e_i = \frac{1}{p-1} \sum_{m=1}^{p-1} \phi^{-i}(m)\, [m].$$

Fix a generator g of G, and set

$$x_i = (g - 1)\, e_i = \frac{1}{p-1} \sum_{m=1}^{p-1} \phi^{-i}(m)\, (g^m - 1).$$

5.1. Theorem. *With the above notation, the Hopf orders in KG are precisely the subrings of the form $\mathfrak{A} = \mathfrak{O}[d^{-1}x_1]$ where $d \in \mathfrak{O}$ is such that the ideal $d^{p-1}\mathfrak{O}$ divides $p\,\mathfrak{O}$. Two values of d give the same Hopf order if and only if they generate the same $\mathfrak{O}$-ideal.*

5.2. Remarks.

(i) : Taking $d = 1$ gives the group ring $\mathfrak{A} = \mathfrak{O}G$, and if K contains a primitive pth root ζ of 1, so $p\mathfrak{O} = (1 - \zeta)^{p-1}\mathfrak{O}$, then taking $d = 1 - \zeta$ gives the split maximal order $\mathfrak{A} = \mathfrak{M}$.

(ii): The number of Hopf orders is governed by the absolute ramification index e of K:- in fact there are $1 + \lfloor e/(p-1) \rfloor$ Hopf orders, where $\lfloor y \rfloor$ denotes the greatest integer not exceeding y. As the $\mathfrak{O}$-ideals are powers of the unique maximal ideal, the lattice of all Hopf orders in KG with respect to inclusion is a line in this local situation.

(iii): If $p > 2$ the generator x_1 can be written in the simpler form

$$x_1 = \frac{1}{p-1} \sum_{m=1}^{p-1} \phi^{-1}(m)\, g^m.$$

PROOF OF THEOREM 5.1: Let $\mathfrak{A}$ be any Hopf order in KG. Then $\mathfrak{A}$ is an $\mathfrak{O}\Lambda$-module, and indeed has a direct sum decomposition as such,

$$(3) \qquad \qquad \mathfrak{A} = \mathfrak{O} \oplus \bigoplus_{i=1}^{p-1} \mathfrak{A}^+ e_i,$$

where $\mathfrak{A}^+$ is the augmentation ideal $\{a \in \mathfrak{A} : \epsilon(a) = 0\}$ of $\mathfrak{A}$. For each i, the summand $\mathfrak{A}^+ e_i$ contains some nonzero multiple of $x_i = (g - 1)\, e_i$, since $\mathfrak{A}K = KG$, and so $\mathfrak{A}^+ e_i$ has $\mathfrak{O}$-rank at least 1. Thus (3) expresses $\mathfrak{O}$ as a direct sum of p modules, each of $\mathfrak{O}$-rank at least 1, and since $\mathfrak{A}$ itself has $\mathfrak{O}$-rank p, it follows that each summand has rank exactly 1.

Let π be a uniformiser of K, and let e be the absolute ramification index (so $\pi^e \mathfrak{O} = p\mathfrak{O}$). Then for each i,

$$\mathfrak{A}^+ e_i = \mathfrak{O}\, \pi^{-a_i}\, x_i$$

for some $a_i \in \mathbb{Z}$. Thus classifying Hopf orders in KG amounts to determining the values of $a_1, \dots, a_{p-1}$ for which the lattice

$$\mathfrak{A} = \mathfrak{O} \oplus \bigoplus_{i=1}^{p-1} \mathfrak{O}\, \pi^{-a_i}\, x_i$$

is closed under both multiplication and comultiplication.

We first consider comultiplication: using the orthogonality relations for the abelian characters ϕ^{-i} of Λ, it is easily verified that

$$\Delta(x_i) = 1 \otimes x_i + x_i \otimes 1 + \sum_{j=1}^{p-1} x_j \otimes x_{i-j}$$

and hence that $\mathfrak{A}$ is costable if and only if

(4) $\qquad a_j + a_{i-j} \geq a_i$ for all i, j.

Here the subscripts are to be interpreted mod $p - 1$.

As for multiplication, a calculation yields

$$(p-1)x_i x_j = \begin{cases} J_{ij}\, x_{i+j} & \text{if } i, j \not\equiv 0 \pmod{p-1} \\ -p\, x_j & \text{if } i \equiv 0 \pmod{p-1} \\ -p\, x_i & \text{if } j \equiv 0 \pmod{p-1}, \end{cases}$$

where J_{ij} is the Jacobi sum

$$J_{ij} = \sum_{m=1}^{p-1} \phi^{-i}(m)\, \phi^{-j}(1-m).$$

One knows that for $1 \leq i,\, j \leq p-2$,

$$J_{ij}\mathfrak{O} = \begin{cases} \mathfrak{O} & \text{if } i+j < p \\ p\mathfrak{O} & \text{if } i+j \geq p \end{cases}$$

(see e.g. [30, Lemma 6.2(d), Proposition 6.13]), and it follows that $\mathfrak{A}$ is closed under multiplication if and only if

(5) $\qquad a_{i+j} \geq \begin{cases} a_i + a_j & \text{if } i+j < p \\ a_i + a_j - e & \text{if } i+j \geq p. \end{cases}$

The conditions (4) and (5) are simultaneously satisfied if and only if $a_i = i\, a_1$ for $1 \leq i \leq p-1$, with $0 \leq a_1 \leq e/(p-1)$. Thus the Hopf orders in A are precisely the rings $\mathfrak{O}[\pi^{-a_1} x_1]$ where the ideal $(\pi^{-a_1}\mathfrak{O})^{p-1}$ divides $p\mathfrak{O}$. Since multiplying the factor $d = \pi^{-a_1}$ by a unit does not change the Hopf order, this completes the proof of the theorem. $\square$

If instead we now take K to be an extension of $\mathbb{Q}_q$ for a prime $q \neq p$ then the group ring $\mathfrak{O}G$ is the maximal order in KG. By Proposition 4.2 every Hopf order contains $\mathfrak{O}G$, whence $\mathfrak{O}G$ is the only Hopf order in KG.

Now take K to be a number field. By the argument of [18, (5.3)], an order in the algebra KG is determined by its local completions, which may be prescribed arbitrarily subject to the constraint that all but finitely many of them are maximal orders. The order is a Hopf order if and only if each of its completions is. Thus we have:

5.3. Corollary. *Let K be a number field and let G be a group of order p. Then the Hopf orders $\mathfrak{A}$ in KG are parameterised by the integral $\mathfrak{O}$-ideals $\mathfrak{d}$ such that $\mathfrak{d}^{p-1}$ divides $p\,\mathfrak{O}$. Explicitly, the Hopf order $\mathfrak{A}(\mathfrak{d})$ corresponding to $\mathfrak{d}$ has completions*

$$\mathfrak{A}(\mathfrak{d})_{\mathfrak{p}} = \mathfrak{O}_{\mathfrak{p}}[d_{\mathfrak{p}}^{-1} x_1]$$

where $d_{\mathfrak{p}}$ is any generator of $\mathfrak{d}\,\mathfrak{O}_{\mathfrak{p}}$ and x_1 is as above.

We now give an ad hoc definition of the conductor of a Hopf order $\mathfrak{A}$ in KG, assuming for simplicity that K contains a primitive pth root ζ of 1.

Let χ be any nontrivial character of G. Then the conductor of $\mathfrak{A}$ is the $\mathfrak{O}$-ideal

$$\mathfrak{f}(\mathfrak{A}) = \{\chi(a) - \epsilon(a) \,:\, a \in \mathfrak{A}\}.$$

This is independent of the choice of χ, since any other nontrivial character has the form $\chi^m = \chi \circ [m]$ with $(m, p) = 1$, and $[m]$ is an automorphism of the Hopf order $\mathfrak{A}$.

5.4. Lemma.

$$\mathfrak{f}(\mathfrak{A}) = (1 - \zeta)\,\mathfrak{d}^{-1}.$$

PROOF: It suffices to show this in the local case, so we may put $\mathfrak{d} = d\,\mathfrak{O}$. Then

$$\mathfrak{f}(\mathfrak{A}) = \chi(\mathfrak{A}^+) = \frac{\chi(x_1)}{d}\,\mathfrak{O}.$$

The case $p = 2$ is easily verified. Taking p odd, and assuming without loss of generality that $\zeta = \chi(g)$, we have

$$
\begin{aligned}
(p-1)\chi(x_1) &= \sum_{m=1}^{p-1} \phi^{-1}(m)\,\zeta^m \\
&\equiv \sum_m m^{-1}(1 - (1-\zeta))^m \quad (\mathrm{mod}\ p\,\mathfrak{O}) \\
&\equiv \sum_m m^{-1}(1 - m(1-\zeta)) \quad (\mathrm{mod}\ (1-\zeta)^2\,\mathfrak{O}) \\
&= 0 - (p-1)(1-\zeta).
\end{aligned}
$$

Thus

$$\mathfrak{f}(\mathfrak{A}) = -\frac{1-\zeta}{d}\,\mathfrak{O} = (1-\zeta)\,\mathfrak{d}^{-1}$$

as required.   □

Combining the lemma with the preceding corollary, we see that for each integral $\mathfrak{O}$-ideal $\mathfrak{f}$ dividing $(1 - \zeta)\,\mathfrak{O}$, there is a unique Hopf order in KG with conductor $\mathfrak{f}$. Its completion at any prime $\mathfrak{p}$ is the unique Hopf order in $K_{\mathfrak{p}}G$

with conductor $\mathfrak{f}\mathfrak{O}_\mathfrak{p}$. The group ring $\mathfrak{O}G$ has conductor $(1-\zeta)\mathfrak{O}$ and the split maximal order has conductor $\mathfrak{O}$.

§6 A Generalisation

If we take G to be an elementary abelian p-group of order $q = p^n > p$, the situation is much more complicated than for groups of order p, and the full classification of Hopf orders in KG is not known, even for $n = 2$. The argument of Theorem 5.1 can however be applied to classify certain special families of Hopf orders in KG, and we will now briefly describe this result. It is the analogue for Hopf orders in a group algebra of a result on group schemes due to Raynaud ([17, Théorème 1.4.1]). For a fuller discussion of the result for Hopf orders, see [2, Theorem 4].

We take K to be a finite extension of $\mathbb{Q}_p$, and assume that K contains a primitive $(q-1)$th root of 1. Let e be the ramification index of K, and let $\mathfrak{p} = \pi\,\mathfrak{O}$ be the maximal ideal of its valuation ring $\mathfrak{O}$.

Let Λ be a totally nonsplit maximal torus in the group $\mathrm{Aut}(G) = GL_n(\mathbb{F}_p)$, i.e. Λ is a subgroup isomorphic to $\mathbb{F}_q^\times$. Identifying these last two groups, we again have a Teichmüller character $\phi : \Lambda \longrightarrow K^\times$, determined by the conditions

$$\phi(\lambda)^{q-1} = 1, \quad \phi(\lambda) \equiv \lambda \pmod{\mathfrak{p}}.$$

The group ring $\mathfrak{O}\Lambda$ has a basis of primitive idempotents

$$e_i = \frac{1}{q-1} \sum_{\lambda \in \Lambda} \phi^{-i}(\lambda)\,\lambda \text{ for } 1 \leq i \leq q-1.$$

For each $\lambda \in \Lambda$, let $[\lambda] : KG \longrightarrow KG$ be the K-linear extension of λ. We consider only those Hopf orders which admit Λ, i.e. for which $[\lambda](\mathfrak{A}) \subseteq \mathfrak{A}$ for all $\lambda \in \Lambda$. Such a Hopf order is an $\mathfrak{O}\Lambda$-module, and, arguing as in the proof of Theorem 5.1, it decomposes as a direct sum

$$\mathfrak{A} = \mathfrak{O} \oplus \bigoplus_{i=1}^{q-1} \mathfrak{O}\pi^{-a_i}\, x_i$$

where

$$x_i = \frac{1}{q-1} \sum_{\lambda \in \Lambda} \phi^{-1}(\lambda)\,(\lambda(g) - 1),$$

$g \neq 1$ being a fixed element of G. The condition on the a_i for this lattice to be closed under comultiplication is (4) as above, but the condition for closure under multiplication is more involved than before, as the arithmetic behaviour of the Jacobi sums is more complicated for $\mathbb{F}_q$ than for $\mathbb{F}_p$. The final result is:

6.1. Theorem. *The Hopf orders $\mathfrak{A}$ in KG which admit Λ are precisely the subrings*

$$\mathfrak{A} = \mathfrak{O}[\pi^{-b_0} x_1, \pi^{-b_1} x_p, \ldots, \pi^{-b_{n-1}} x_{p^{n-1}}]$$

where the integers $b_0, b_1, \ldots, b_{n-1}$ satisfy the inequalities

$$p\,b_i - e \leq b_{i+1} \leq p\,b_i \quad \text{for } 0 \leq i \leq n - 1.$$

(The subscripts on the b_i are to be read mod n.)

§7   Group Schemes and Hopf Algebras

Let K be any commutative ring. Recall from algebraic geometry (see for instance [10]) that to each commutative K-algebra R is associated the affine scheme $\mathrm{Spec}(R)$ over K. This is by definition the ringed space $(X, \mathcal{O}_X)$, where X is the topological space consisting of the prime ideals of R, with closed sets the subsets $V(I) = \{\mathfrak{p} \in X : I \subseteq \mathfrak{p}\}$ for all ideals I of R, and where $\mathcal{O}_X$ is the sheaf of K-algebras on X associating to each open set $X \backslash V(I)$ of X the localisation $R_{(I)}$ of R at I. Morphisms $f : \mathrm{Spec}(S) \longrightarrow \mathrm{Spec}(R)$ of affine schemes correspond to homomorphisms $\widehat{f} : R \longrightarrow S$ of K-algebras.

Equivalently, the affine scheme $\mathrm{Spec}(R)$ can be viewed as the functor associating to every commutative K-algebra S the set $\mathrm{Hom}_{K-\mathrm{alg}}(R, S)$ of K-algebra homomorphisms from R to S. For example, $\mathrm{Spec}(K[T])$ is the affine line over K; in particular, if K is a field then the functor $\mathrm{Spec}(K[T])$ associates to every field extension L of K the set of L-rational points on the affine line, i.e. a copy of the field L.

An *affine group scheme* $\mathcal{G} = \mathrm{Spec}(A)$ is an affine scheme together with morphisms of schemes

$$
\begin{array}{rcll}
m & : & \mathcal{G} \times \mathcal{G} \longrightarrow \mathcal{G} & \text{(multiplication)} \\
u & : & \mathrm{Spec}(K) \longrightarrow \mathcal{G} & \text{(inclusion of unit element)} \\
\iota & : & \mathcal{G} \longrightarrow \mathcal{G} & \text{(inverse)}
\end{array}
$$

satisfying the commutative diagrams expressing the usual group axioms (cf. [31, p.7]). In terms of K-algebras, this means that A is equipped with algebra homomorphisms

$$
\begin{array}{rcll}
\Delta = \widehat{m} & : & A \longrightarrow A \otimes_K A \\
\epsilon = \widehat{u} & : & A \longrightarrow K \\
{}^- = \widehat{\iota} & : & A \longrightarrow A
\end{array}
$$

such that

$$
\begin{array}{rcll}
(1 \otimes \Delta) \circ \Delta = (\Delta \otimes 1) \circ \Delta & : & A \longrightarrow A \otimes A \otimes A \\
(\epsilon \otimes 1) \circ \Delta = 1 = (1 \otimes \epsilon) \circ \Delta & : & A \longrightarrow K \otimes A \cong A \cong A \otimes K \\
m_A \circ ({}^- \otimes 1) \circ \Delta = i \circ \epsilon = m_A \circ (1 \otimes {}^-) \circ \Delta & : & A \longrightarrow A
\end{array}
$$

where $i : K \longrightarrow A$ is the K-linear map determined by $1_K \mapsto 1_A$.

We call a commutative algebra A equipped with such maps a (commutative) *Hopf algebra* over K. If G is a finite abelian group, it is easily verified that the algebras $A = KG$ and $B = \mathrm{Map}(K, G)$ discussed earlier in this chapter are Hopf algebras over K.

Now suppose that K is the field of fractions of a Dedekind domain $\mathfrak{O}$ of characteristic 0. For each prime ideal $\mathfrak{p}$ of $\mathfrak{O}$ we write $\mathfrak{O}_{(\mathfrak{p})}$ for the localisation of $\mathfrak{O}$ at $\mathfrak{p}$ and $k(\mathfrak{p})$ for the residue field $\mathfrak{O}_{(\mathfrak{p})}/\mathfrak{p}\,\mathfrak{O}_{(\mathfrak{p})}$ at $\mathfrak{p}$. In particular, for the generic prime $\mathfrak{g} = (0)$ we have $\mathfrak{O}_{(\mathfrak{g})} = k(\mathfrak{g}) = K$.

For any finite dimensional commutative Hopf algebra A over K, we call an order $\mathfrak{A}$ in A a *Hopf order* if it is costable with respect to the comultiplication map Δ, as before. Then $\mathcal{G} = \mathrm{Spec}(\mathfrak{A})$ is a group scheme over $\mathfrak{O}$. Since $\mathfrak{A}$ is finitely generated and torsion-free as an $\mathfrak{O}$-module, $\mathcal{G}$ is finite and flat as a group scheme. For each $\mathfrak{p}$, the fibre $\mathcal{G}_{\mathfrak{p}}$ of $\mathcal{G}$ at $\mathfrak{p}$ is by definition the affine group scheme $\mathrm{Spec}(\mathfrak{A} \otimes_{\mathfrak{O}} k(\mathfrak{p}))$ over $k(\mathfrak{p})$. In particular, the generic fibre $\mathcal{G}_{\mathfrak{g}}$ is the group scheme $\mathrm{Spec}(A)$ over K. Conversely, any finite flat affine group scheme over $\mathfrak{O}$ whose generic fibre is $\mathrm{Spec}(A)$ is given by $\mathrm{Spec}(\mathfrak{A})$ for some Hopf order $\mathfrak{A}$ in A.

We may think of a Hopf order $\mathfrak{A}$ in A as determining a continuously varying family of group schemes

$$\{\mathrm{Spec}(\mathfrak{A})_{\mathfrak{p}} \text{ a group scheme over } k(\mathfrak{p}) : \mathfrak{p} \in \mathrm{Spec}(\mathfrak{O})\}$$

with $\mathrm{Spec}(\mathfrak{A})_{\mathfrak{g}}$ coinciding with $\mathrm{Spec}(A)$. In particular, if $\mathfrak{O}$ is local with unique maximal ideal $\mathfrak{p}$, this family consists of just two members, the generic fibre $\mathrm{Spec}(A)$ and the special fibre $\mathrm{Spec}(\mathfrak{A}_{\mathfrak{p}}) = \mathrm{Spec}(\mathfrak{A} \otimes_{\mathfrak{O}} \mathfrak{O}/\mathfrak{p})$.

As an example, we will show how Hopf orders can be constructed from formal groups. Let K be a finite extension of the p-adic field $\mathbb{Q}_p$, let $\mathfrak{O}$ be its valuation ring, and let $\mathfrak{p}$ be the unique maximal ideal of $\mathfrak{O}$. Recall (cf. [19]) that a *commutative (1-dimensional) formal group* over $\mathfrak{O}$ is a formal power series $F(X, Y) \in \mathfrak{O}[[X, Y]]$ such that

(1) $F(0, Y) = Y$ (left identity);
(2) $F(X, Y) = F(Y, X)$ (commutativity);
(3) $F(F(X, Y), Z) = F(X, F(Y, Z))$ (associativity);
(4)   there exists $G(X) \in \mathfrak{O}[[X]]$ such that $F(X, G(X)) = 0$ (inverse);
(5) $F(X, Y) \equiv X + Y \bmod \deg 2$;

where two power series are said to be congruent mod deg 2 if they agree in all terms of total degree less than 2. In fact, the conditions (4) and (5) are consequences of (1), (2) and (3).

If L is any extension of K, and if $\mathfrak{p}_L$ is the maximal ideal of the valuation ring $\mathfrak{O}_L$ of L, then we can view $\mathfrak{p}_L$ as a group by means of $F(X, Y)$:- explicitly, for x, $y \in \mathfrak{p}_L$ we set $x +_F y = F(x, y)$, which is clearly convergent to an element of $\mathfrak{p}_L$.

Let $\mathfrak{O}[[X]] \,\widehat{\otimes}\, \mathfrak{O}[[X]]$ denote the completion of the tensor product $\mathfrak{O}[[X]] \otimes \mathfrak{O}[[X]]$ with respect to the ideal generated by the elements $Y = X \otimes 1$ and $Z = 1 \otimes X$. Thus we may identify $\mathfrak{O}[[X]] \,\widehat{\otimes}\, \mathfrak{O}[[X]]$ with $\mathfrak{O}[[Y, Z]]$.

Let $\Delta : \mathfrak{O}[[X]] \longrightarrow \mathfrak{O}[[X]] \,\widehat{\otimes}\, \mathfrak{O}[[X]]$, $\epsilon : \mathfrak{O}[[X]] \longrightarrow \mathfrak{O}$ and $^{-} : \mathfrak{O}[[X]] \longrightarrow \mathfrak{O}[[X]]$ be the homomorphisms of $\mathfrak{O}$-algebras determined by $\Delta(X) = F(Y, Z)$, $\epsilon(X) = 0$ and $\overline{X} = G(X)$. This makes $\mathfrak{O}[[X]]$ into a "topological" Hopf algebra, defined as above except that the comultiplication takes $\mathfrak{O}[[X]]$ into the completed tensor product rather than the usual algebraic tensor product.

A power series $h(X) \in \mathfrak{O}[[X]]$ is an endomorphism of the formal group F if $h(F(X, Y)) = F(h(X), h(Y))$. For any endomorphism $h(X)$, the quotient ring

$$R = \frac{\mathfrak{O}[[X]]}{(h(X))}$$

inherits the structure of a topological Hopf algebra over $\mathfrak{O}$. We now restrict attention to those $h(X)$ of the following form:-

$$h(X) = u(X)\,\tilde{h}(X).$$

where $u(X)$ is a unit in $\mathfrak{O}[[X]]$ and where $\tilde{h}(X)$ is a monic polynomial which splits over K, and all of whose roots lie in $\mathfrak{p}$. Let n denote the degree of $\tilde{h}$. Then R is a torsion-free $\mathfrak{O}$-algebra of rank n, so $R\widehat{\otimes}R$ can be identified with $R \otimes R$. Moreover, the roots of $\tilde{h}$ form an abelian group G of order n under $+_F$. Evaluating an element $r(X)$ mod $(h(X))$ of R at any $g \in G$ gives a well-defined element of $\mathfrak{O}$, so we may identify the Hopf $\mathfrak{O}$-algebra R with a subring of $\mathrm{Map}(G, \mathfrak{O})$. We have thus produced a Hopf order R in the Hopf algebra $B = \mathrm{Map}(G, K)$.

For any commutative $\mathfrak{O}$-algebra S, the formal group F gives a group structure to the set $G_S = \{s \in S : \tilde{h}(s) = 0\}$. The group scheme $\mathrm{Spec}(R)$ can be interpreted as the functor taking each S to the group G_S.

We remark that by placing somewhat stronger conditions on the power series $h(X)$, namely that it be a Lubin-Tate series for $\mathfrak{O}$ (cf. [19]), we can ensure the existence of a unique formal group over $\mathfrak{O}$ for which h is an endomorphism.

Another source of formal groups is elliptic curves. In this case, the formal group law expresses the group law on the elliptic curve in a neighbourhood of the origin. We shall use elliptic curves to generate Hopf orders in Chapter 5.

III Principal Homogeneous Spaces

As before, let K be the field of fractions of a Dedekind domain $\mathfrak{O}$ of character-istic 0, and let G be a finite abelian group. We will continue to work with Hopf orders in $A = KG$ and $B = \mathrm{Map}(G, K)$. In this chapter we will be concerned with the objects on which a Hopf order $\mathfrak{A}$ in A acts. Rather than studying all $\mathfrak{A}$-modules, we will make use of the comultiplication in $\mathfrak{A}$ by considering only those $\mathfrak{A}$-modules which have the structure of an $\mathfrak{O}$-algebra, and are in fact "twisted" versions of $\mathfrak{B} = \mathfrak{A}^D$. These objects are the principal homogeneous spaces for the Hopf order $\mathfrak{B}$, and the set of isomorphism classes of principal homogeneous spaces can be given the structure of an abelian group $\mathrm{PH}(\mathfrak{B})$. As in the previous chapter, we will first work at the level of K-algebras, and then see how the theory lifts to integral level. We shall then construct a group homomorphism ψ from $\mathrm{PH}(\mathfrak{B})$ to the locally free classgroup $\mathrm{Cl}(\mathfrak{A})$. Finally, in the case that G is cyclic of order p and K contains a primitive pth root of unity, we use Kummer theory to give an explicit description of $\mathrm{PH}(\mathfrak{B})$ and of the kernel of ψ.

Much of the machinery developed in this chapter has its origins in McCulloh's work [14] on the determination of the classes in $\mathrm{Cl}(\mathfrak{O}G)$ which are realisable by tame rings of integers. McCulloh does not, however, make explicit use of Hopf algebra theory, since he is concerned only with the case $\mathfrak{A} = \mathfrak{O}G$.

§1 Principal Homogeneous Spaces for B

Recall that G acts on B by the rule $f^g(h) = f(hg^{-1})$ for g, $h \in G$, $f \in B$. Thus each $g \in G$ induces an algebra automorphism on B. Moreover, $f^g = f$ for all $g \in G$ if and only if f is the constant map $k\,1_B$ for some $k \in K$. Identifying $K\,1_B \subseteq B$ with K, we therefore have $B^G = K$.

This situation is analogous to that considered in Galois theory, but with a finite dimensional commutative K-algebra B in place of a field extension of K. With this in mind, we define the category of G-*Galois algebras over K* as follows:-

- its objects are the K-algebras C on which G acts as algebra automor-phisms and for which there exists a field extension L of K and a G-equivariant isomorphism

$$\xi : C \otimes_K L \longrightarrow B \otimes_K L = \mathrm{Map}(G, L)$$

of L-algebras. We call ξ a *splitting isomorphism*, and L a *splitting field*, for C.

- its morphisms are G-equivariant isomorphisms $\phi : C \longrightarrow C'$ of K-algebras (so every morphism is in fact an isomorphism).

We call an object C in this category a *principal homogeneous space* for B. Obviously B itself is a principal homogeneous space for B. We can easily classify the principal homogeneous spaces for B:

1.1. Proposition. *Let M/K be a Galois extension of fields whose Galois group is identified with a subgroup H of G. Then*

$$C_M = \mathrm{Map}_H(G, M)$$

is a principal homogeneous space for B. Conversely, up to isomorphism, every principal homogeneous space for B has this form.

PROOF: C_M is a K-algebra under pointwise addition and multiplication of maps, and G acts on it as algebra automorphisms via the rule $f^g(g') = f(g' g^{-1})$ for $f \in C_M$ and $g, g' \in G$. Let n be the order of G, and s the index of H in G. Any element of C_M is determined by its values on a system of coset representatives $g_1, \ldots, g_s$ of H in G, and these values may be chosen arbitrarily in M. Thus as an algebra, C_M is the product of s copies of M, and so has dimension n over K. Taking M itself as a splitting field, we will show that the G-equivariant homomorphism of M-algebras

$$\pi : \mathrm{Map}_H(G, M) \otimes M \longrightarrow \mathrm{Map}(G, M),$$

given by

$$\pi(f \otimes m)(g) = f(g)\, m,$$

is an isomorphism. Comparing dimensions, it is enough to show that π is injective. For this, take a basis $m_1, \ldots, m_r$ of M/K, so $rs = n$, and for $1 \leq i \leq r$, let c_i be the element of C_M defined by

$$c_i(g) = \begin{cases} m_i{}^g & \text{if } g \in H \\ 0 & \text{if } g \in G \backslash H. \end{cases}$$

Then $\{c_i{}^{g_j} : 1 \leq i \leq r, 1 \leq j \leq s\}$ is a K-basis for C_M. Suppose that, for some $a_{ijk} \in K$, we have

$$\pi\left(\sum_{i,k=1}^{r} \sum_{j=1}^{s} a_{ijk}\, c_i{}^{g_j} \otimes m_k \right) = 0.$$

Evaluating at hg_j and applying $y \in H$, it follows that

$$\sum_{i,k=1}^{r} a_{ijk}\, m_i{}^{hy}\, m_k{}^{y} = 0$$

for each j and for all $h, y \in H$. Now the matrix $(m_i{}^y)$ is nonsingular, since M/K is separable. Thus each coefficient a_{ijk} is zero, which shows that π is indeed injective, and hence is a splitting isomorphism for C_M.

Conversely, let C be a principal homogeneous space for B, and let L be a splitting field. Then

$$C \otimes L = \prod_1^n L$$

as an L-algebra, where n is the order of G, so C must be a product $\prod M_i$ of subfields M_i of L. Let e be the primitive idempotent of C corresponding to the field component $M = M_1$, and let H be the subgroup of G defined by

$$H = \{ g \in G \ : \ e^g = e \}.$$

Now G acts transitively on the primitive idempotents of $C \otimes L$, and hence also on the primitive idempotents of C. Thus the distinct primitive idempotents of C are the e^g, and the distinct field components M_i of C are the M^g, where g runs through a transversal of H in G. Hence, writing s for the index of H in G as before, C is the product of s fields, each isomorphic to M.

Each element h of H induces a field automorphism of M, fixing K elementwise, and this automorphism is the identity only if $h = 1$ in G. Moreover, as C has dimension n over K, it follows that M/K has degree $r = n/s = |H|$. This shows that M/K is a Galois extension with group H, and identifying the members of a basis $m_1, \ldots, m_r$ of M/K with maps $c_1, \ldots, c_r$ as above, we have shown that, up to isomorphism, $C = C_M$. □

Taking $M = K$ in the Proposition, so that $H = \{1\}$, we obtain $C_K = B$. At the other extreme, if $\mathrm{Gal}(M/K) \cong G$ then C_M is just the field M. Intuitively, what we have done is to enlarge the family of Galois extensions M/K with group isomorphic to G, so as to allow also degenerate cases with Galois group isomorphic to a subgroup of G (and in particular, to allow K itself), all the time working with K-algebras of dimension $|G|$. We will see below that this enables us to put a group structure on the set of (isomorphism classes of) such extensions.

This approach, which was originated by Grothendieck, gives an alternative to the usual formulation (in terms of fixed points of groups of automorphisms) of the Galois theory of fields. The new approach extends to the integral level, allowing us to do Galois theory for Hopf orders. We will discuss this more fully in the next section.

We next determine to what extent the splitting isomorphisms ξ are unique.

1.2. Lemma. *The group* $\mathrm{Aut}_{G-\mathrm{Gal}/K}(B)$ *of automorphisms of B as a G-Galois algebra is precisely G.*

PROOF: If ψ is such an automorphism, it must take primitive idempotents to primitive idempotents. Thus, writing ℓ for the element of B defined by

$$(6) \qquad\qquad \ell(g) = \begin{cases} 1 & \text{if } g = 1 \\ 0 & \text{otherwise,} \end{cases}$$

as before, we have $\psi(\ell) = \ell^g$ for some $g \in G$. Then

$$\psi(\ell^h) = \psi(\ell)^h = \ell^{gh} = \ell^{hg}$$

for all $h \in G$, and the automorphism ψ is that induced by g.   □

1.3. Lemma. *If C is a principal homogeneous space for B then any splitting isomorphism* $\xi : C \otimes L \longrightarrow B \otimes L$ *has the form*

$$\xi(c \otimes m)(g) = \tau(c^{g^{-1}})\,m \quad \text{for some algebra homomorphism } \tau : C \longrightarrow L.$$

PROOF: By Proposition 1.1, $C = C_M$ for some subfield M of L, Galois over K, and of degree r say. Let n be the order of G, and let $s = n/r$. Then, as a K-algebra,

$$C \cong \prod_1^s M,$$

and there are precisely n algebra homomorphisms $\tau : C \longrightarrow L$, namely the composites

$$C \xrightarrow{p} M \xrightarrow{\sigma} L,$$

where p is one of the s projections onto a factor M, and σ is one of the r K-linear embeddings of M into L. The corresponding maps ξ are clearly distinct and, arguing as in Proposition 1.1, are indeed splitting isomorphisms for C.

It remains to show that there can be no more than n such isomorphisms. Now if ζ and ζ' are two splitting isomorphisms for C we may assume (by extending scalars if necessary) that the splitting field L is the same in both cases. Then $\zeta' \circ \zeta^{-1}$ is an automorphism of $B \otimes_K L$ as a G-Galois algebra over L. By Lemma 1.2 with L in place of K, there are precisely n such automorphisms, and hence, keeping ζ fixed, precisely n splitting isomorphisms ζ' for C.   □

We write $\mathrm{PH}(B)$ for the set of isomorphism classes of principal homogeneous spaces for B. We can obtain a cohomological interpretation of this set by the general method of Galois descent (see e.g. [20, X §2]). Writing K^c for a fixed algebraic closure of K, we may view the elements of $\mathrm{PH}(B)$ as the isomorphism classes of K-algebras-with-G-action which, on extension of the base field from K to K^c, give $B \otimes K^c$. Let $\Omega = \mathrm{Gal}(K^c/K)$ be the absolute Galois group of K, and let A be the group of automorphisms of $B \otimes K^c$ as a principal homogeneous space for itself:- by Lemma 1.2, A is isomorphic to G, and we write α_g for the element of A corresponding to $g \in G$. Then Ω acts on A by the rule $\alpha_g{}^\omega = \omega \circ \alpha_g \circ \omega^{-1}$. Since α_g merely permutes the factors K^c in $B \otimes K^c = \prod K^c$, this action is trivial, and thus we have

$$(7) \qquad\qquad \alpha \circ \omega = \omega \circ \alpha \quad \text{for all } \alpha \in A,\ \omega \in \Omega.$$

Now let C be a principal homogeneous space for B, with splitting isomorphism

$$\xi : C \otimes K^c \longrightarrow B \otimes K^c.$$

For any $\omega \in \Omega$, set

$$\xi_\omega = \xi \circ \omega \circ \xi^{-1} \circ \omega^{-1} \in A.$$

Then, using (7), it easily follows that $\xi_\omega \circ \xi_{\omega'} = \xi_{\omega\omega'}$ for any $\omega, \omega' \in \Omega$. Moreover, if ξ' is any other splitting isomorphism for C then $\xi' \circ \xi^{-1} \in A$, and it follows from (7) that $\xi'_\omega = \xi_\omega$ for all $\omega \in \Omega$. Associating to C the map $\omega \mapsto \xi_\omega$, we have therefore constructed a map

$$\mathrm{PH}(B) \longrightarrow \mathrm{Hom}(\Omega, A) = H^1(\Omega, A) \cong H^1(\Omega, G).$$

Arguing as in the proof of [20, X §2 Proposition 4], it can be shown that it is in fact a bijection. The group structure of $\mathrm{Hom}(\Omega, G)$ then gives rise to a natural abelian group structure on $\mathrm{PH}(B)$.

1.4. Example. If G is cyclic of order n and if K contains a primitive nth root of 1, then by Kummer theory we have

$$\mathrm{Hom}(\Omega, G) \cong \frac{K^\times}{(K^\times)^n}.$$

Thus, fixing an identification $g \leftrightarrow \zeta_g$ of G with the group of nth roots of 1 in K, we have the explicit isomorphism

$$\mathrm{PH}(B) \cong \frac{K^\times}{(K^\times)^n}$$

where the class $a \bmod (K^\times)^n$ corresponds to

$$\frac{K[X]}{(X^n - a)} \quad \text{with } G\text{-action } X^g = X\zeta_g.$$

$\square$

We next give an explicit description of the multiplication in $\mathrm{PH}(B)$. For this we introduce the diagonal subgroup

$$\delta(G) = \{(g, g^{-1}) \ : \ g \in G\} \subseteq G \times G.$$

1.5. Lemma. *Let C and D be two principal homogeneous spaces for B. Then the product of their classes in $\mathrm{PH}(B)$ is represented by*

$$C.D = (C \otimes_K D)^{\delta(G)}.$$

(Here we are taking the $\delta(G)$-fixed points of $C \otimes D$, considered as a $G \times G$-module in the obvious fashion.)

PROOF: We first verify that $C.D$ is a principal homogeneous space for B. It is a K-algebra of dimension $|G|$, and has a well-defined action of G (via either tensor factor) as algebra automorphisms. Moreover we have

$$
\begin{aligned}
(C.D) \otimes K^c &= (C \otimes D)^{\delta(G)} \otimes K^c \\
&\cong (C \otimes B \otimes K^c)^{\delta(G)} \\
&\cong (B \otimes B \otimes K^c)^{\delta(G)} \\
&= (B \otimes B)^{\delta(G)} \otimes K^c \\
&= \Delta(B) \otimes K^c \\
&\cong B \otimes K^c.
\end{aligned}
$$

Explicitly, if $\xi^{(C)}$ and $\xi^{(D)}$ are splitting isomorphisms for C and D respectively, then $C.D$ has a splitting isomorphism

$$
\xi^{(C.D)} = \Delta^{-1} \circ (\xi^{(C)} \otimes \xi^{(D)}) : C.D \otimes K^c = ((C \otimes K^c) \otimes_{K^c} (D \otimes K^c))^{\delta(G)} \longrightarrow B \otimes K^c.
$$

Since the action of Ω commutes with $\Delta : B \otimes K^c \longrightarrow B \otimes B \otimes K^c$, we have

$$
\begin{aligned}
(\xi^{(C.D)})_\omega &= \Delta^{-1} \circ (\xi^{(C)} \otimes \xi^{(D)}) \circ \omega \circ (\Delta^{-1} \circ (\xi^{(C)} \otimes \xi^{(D)}))^{-1} \circ \omega^{-1} \\
&= \Delta^{-1} \circ ((\xi^{(C)} \otimes \xi^{(D)}) \circ \omega \circ (\xi^{(C)} \otimes \xi^{(D)})^{-1} \circ \omega^{-1}) \circ \Delta \\
&= \Delta^{-1} \circ ((\xi^{(C)}_\omega \otimes 1) \circ (1 \otimes \xi^{(D)}_\omega)) \circ \Delta,
\end{aligned}
$$

so, as automorphisms of $B \otimes K^c$, we have $\xi^{(C.D)}_\omega = \xi^{(C)}_\omega \xi^{(D)}_\omega$. Thus $C.D$ is indeed the product of C and D.   □

One can of course verify, without using Galois cohomology, that the above construction does define an abelian group structure on $\mathrm{PH}(B)$. The first part of the proof of Lemma 1.5 shows that $\mathrm{PH}(B)$ is closed under this operation (we can replace K^c by a finite extension L of K which splits both C and D). We will now outline the remaining steps of this verification.

Associativity: If C, D, E are three principal homogeneous spaces then

$$
(C.D).E = ((C \otimes D)^{\delta(G)} \otimes E)^{\delta(G)} = (C \otimes D \otimes E)^{\kappa(G)}
$$

where $\kappa(G)$ is the subgroup of $G \times G \times G$ generated by all elements of the form $(g, g^{-1}, 1)$ or $(h, 1, h^{-1})$. But this is precisely the kernel of the multiplication map $G \times G \times G \longrightarrow G$, so by symmetry we have $C.(D.E) = (C \otimes D \otimes E)^{\kappa(G)}$.

Commutativity: This is clear from the construction.

Identity: B is the identity of $\mathrm{PH}(B)$.

PROOF: Let $\ell \in B = \mathrm{Map}(G, K)$ be the element defined by (6). If C is any principal homogeneous space for B, then $C.B = (C \otimes B)^{\delta(G)}$ consists of elements

$$
\sum_{h \in G} c_h \otimes \ell^h,
$$

where $c_{h-1} = c_1{}^h$ for all h. It easily follows that $C.B \cong C$.   □

Inverse: The inverse of the class of C is the class of the principal homogeneous space C^*, whose underlying algebra is the same of that of C (we write c^* for the element of C^* corresponding to $c \in C$) and on which G acts by the rule $(c^*)^g = (c^{g^{-1}})^*$.

PROOF: Let L be a splitting field for C, and let ξ be a splitting isomorphism. By Lemma 1.3, there is an algebra homomorphism $\tau : C \longrightarrow L$ such that

$$\xi(c \otimes m)(g) = \tau(c^{g^{-1}}) m$$

for $c \in C$, $m \in L$ and $g \in G$. Then C^* has a splitting isomorphism ξ^* given by

$$\xi^*(c^* \otimes m)(g) = \tau((c^*)^{g^{-1}}) m = \tau(c^g) m.$$

Writing $\mathrm{Tr}_{\delta(G)} : C \otimes C^* \longrightarrow C \otimes C^*$ for the trace map defined by

$$\mathrm{Tr}_{\delta(G)}(c \otimes d^*) = \sum_{(g,g^{-1}) \in \delta(G)} (c \otimes d^*)^{(g,g^{-1})} = \sum_{g \in G} c^g \otimes (d^g)^*,$$

we see that $C.C^* = \mathrm{Tr}_{\delta(G)}(C \otimes C^*)$. Now for any $c \in C$, $d^* \in C^*$, $m \in L$ and $g \in G$ we have

$$
\begin{aligned}
(\xi \otimes \xi^*)(\mathrm{Tr}_{\delta(G)}(c \otimes d^*) \otimes m)(g) &= (\xi \otimes \xi^*)(\sum_{h \in G}(c^h \otimes (d^h)^*) \otimes m)(g) \\
&= \sum_{h \in G} \xi(c^h \otimes 1)(g) \otimes \xi^*((d^h)^* \otimes m)(g) \\
&= \tau(\sum_{h \in G} c^{hg^{-1}} d^{hg}) m.
\end{aligned}
$$

Since

$$\sum_{h \in G} c^{hg^{-1}} d^{hg} = \sum_{h \in G} \left(c^{g^{-1}} d^g \right)^h \in K,$$

this shows that $\xi \otimes \xi^* : C \otimes C^* \otimes L \longrightarrow B \otimes L$ restricts to an isomorphism $C.C^* \longrightarrow \mathrm{Map}(G, K) = B$, so C^* does indeed represent the inverse class to that of C.   □

§2 Principal Homogeneous Spaces for $\mathfrak{B}$

Now let $\mathfrak{A}$ be a Hopf order in A and let $\mathfrak{B} = \mathfrak{A}^D$ be its dual Hopf order in B. A *principal homogeneous space* for $\mathfrak{B}$ is an $\mathfrak{O}$-algebra $\mathfrak{C}$ on which G acts, such that $\mathfrak{C}$ is an order in some principal homogeneous space $C = \mathfrak{C}K$ for B, and such that the splitting isomorphism ξ for C induces an isomorphism

$$\xi : \mathfrak{C} \otimes \mathfrak{O}_L \longrightarrow \mathfrak{B} \otimes \mathfrak{O}_L.$$

(By Lemma 1.2, this condition is independent of the choice of ξ.) As ξ is G-equivariant and $\mathfrak{B}$ admits $\mathfrak{A}$, it follows that $\mathfrak{C}$ is also an $\mathfrak{A}$-module.

We write $\mathrm{PH}(\mathfrak{B})$ for the set of all isomorphism classes of principal homogeneous spaces for $\mathfrak{B}$. This again has the structure of an abelian group, since the construction outlined above carries over to the integral level:- the fact that $\mathfrak{B}$ is a Hopf order is needed to ensure that $\Delta(\mathfrak{B}) = (\mathfrak{B} \otimes \mathfrak{B})^{\delta(G)}$ in proving closure, and to ensure that $\mathfrak{C}.\mathfrak{C}^* \otimes \mathfrak{O}_L \cong \mathfrak{B} \otimes \mathfrak{O}_L$ in showing the existence of inverses.

We now compare $\mathrm{PH}(\mathfrak{B})$ and $\mathrm{PH}(B)$:

2.1. Lemma. *The map* $\mathrm{PH}(\mathfrak{B}) \longrightarrow \mathrm{PH}(B)$ *given by* $\mathfrak{C} \mapsto \mathfrak{C}K$ *is an injective group homomorphism.*

PROOF: It is clear from the definitions of the group operations that the map is a homomorphism. If $\mathfrak{C}$ and $\mathfrak{C}'$ are two principal homogeneous spaces for $\mathfrak{B}$ which are orders in the same principal homogeneous space C for B, then $\mathfrak{C} \otimes \mathfrak{O}_L$ and $\mathfrak{C}' \otimes \mathfrak{O}_L$ have the same image under ξ. Since $\mathfrak{O}_L$ is free over $\mathfrak{O}$, this implies that $\mathfrak{C} = \mathfrak{C}'$. $\qquad\square$

The map in the lemma is in general far from being surjective, as the next example illustrates:

2.2. Example. *Let* $\mathfrak{A} = \mathfrak{O}G$, *so* $\mathfrak{B} = \mathrm{Map}(G, \mathfrak{O})$ *is the maximal order in* B. *Then the principal homogeneous spaces for* $\mathfrak{B}$ *are the maximal orders in the* unramified G-*Galois algebras* C *over* K.

PROOF: Let $\mathfrak{C}$ be a principal homogeneous space for $\mathfrak{B}$. Since $\mathfrak{B}$ has discriminant $\mathfrak{d}(\mathfrak{B}) = \mathfrak{O}$ and since $\mathfrak{C} \otimes \mathfrak{O}_L \cong \mathfrak{B} \otimes \mathfrak{O}_L$, we have $\mathfrak{d}(\mathfrak{C})\,\mathfrak{O}_L = \mathfrak{d}(\mathfrak{B})\,\mathfrak{O}_L = \mathfrak{O}_L$, and hence $\mathfrak{d}(\mathfrak{C}) = \mathfrak{O}$. This occurs precisely when $\mathfrak{C}$ is the maximal order in an unramified algebra. $\qquad\square$

This example provides a particular case where the principal homogeneous space $\mathfrak{C}$ is the integral closure $\mathfrak{O}_C$ of $\mathfrak{O}$ in the algebra $C = \mathfrak{C}K$. Although this sometimes occurs for nonmaximal Hopf orders $\mathfrak{B}$, the relationship between $\mathfrak{C}$ and $\mathfrak{O}_C$ is in general more subtle. The Galois module structure of $\mathfrak{C}$ does, however, provide important information about the Galois module structure of $\mathfrak{O}_C$. In fact, as the next result shows, $\mathfrak{C}$ is always the largest $\mathfrak{A}$-module in $\mathfrak{O}_C$. This was first shown, in a somewhat different context (and by a different method), in [27].

2.3. Proposition. *If* $\mathfrak{C} \in \mathrm{PH}(\mathfrak{B})$ *then* $\mathfrak{C}$ *coincides with*

$$\widetilde{\mathfrak{C}} = \{c \in \mathfrak{O}_C \ : \ c\alpha \in \mathfrak{O}_C \text{ for all } \alpha \in \mathfrak{A}.\}$$

PROOF: Since $\widetilde{\mathfrak{C}}$ is the largest $\mathfrak{A}$-module contained in $\mathfrak{O}_C$, we certainly have $\mathfrak{C} \subseteq \widetilde{\mathfrak{C}}$. It is sufficient to prove the reverse inclusion.

The splitting isomorphism $\xi : \mathfrak{C} \otimes \mathfrak{O}_L \longrightarrow \mathfrak{B} \otimes \mathfrak{O}_L$ extends to an embedding

$$\tilde{\xi} : \tilde{\mathfrak{C}} \otimes_{\mathfrak{O}} \mathfrak{O}_L \longrightarrow B \otimes_K L = \operatorname{Hom}_K(A, L).$$

By Lemma 1.3, $\tilde{\xi}(c \otimes m)(\alpha) = \tau(c\bar{\alpha})m$ for all $c \in \tilde{\mathfrak{C}}$, $m \in \mathfrak{O}_L$ and $\alpha \in A$, where $\tau : C \longrightarrow L$ is some homomorphism of K-algebras. Now if $\alpha \in \mathfrak{A}$ then also $\bar{\alpha} \in \mathfrak{A}$, since $\mathfrak{A}$, being a Hopf order, admits the antipode. Thus for any $c \in \tilde{\mathfrak{C}}$ we have $c\bar{\alpha} \in \mathfrak{O}_C$, and hence $\tau(c\bar{\alpha}) \in \mathfrak{O}_L$. This shows that

$$\tilde{\xi}(\tilde{\mathfrak{C}} \otimes \mathfrak{O}_L) \subseteq \operatorname{Hom}_{\mathfrak{O}}(\mathfrak{A}, \mathfrak{O}_L) = \mathfrak{B} \otimes \mathfrak{O}_L = \tilde{\xi}(\mathfrak{C} \otimes \mathfrak{O}_L).$$

Since $\tilde{\xi}$ is injective, it follows that $\tilde{\mathfrak{C}} \otimes \mathfrak{O}_L \subseteq \mathfrak{C} \otimes \mathfrak{O}_L$, so that $\tilde{\mathfrak{C}} \subseteq \mathfrak{C}$, as required. $\qquad\square$

§3 Class Invariants

The whole point of introducing principal homogeneous spaces for $\mathfrak{B}$ is the following:

3.1. Proposition. *If $\mathfrak{C}$ is a principal homogeneous space for $\mathfrak{B}$ then $\mathfrak{C}$ is a locally free $\mathfrak{A}$-module.*

PROOF: By Theorem 4.3 of Chapter II, $\mathfrak{B}$ is locally free, and hence projective, as an $\mathfrak{A}$-module, so the same is true of $\mathfrak{B} \otimes \mathfrak{O}_L$ and therefore also of $\mathfrak{C} \otimes \mathfrak{O}_L$. But $\mathfrak{C}$ is an $\mathfrak{A}$-module direct summand of $\mathfrak{C} \otimes \mathfrak{O}_L$ and it follows that $\mathfrak{C}$ is $\mathfrak{A}$-projective. Moreover, the A-module $C = \mathfrak{C}K$ affords the regular character of G, since $C \otimes L \cong B \otimes L$ as LG-modules, so C is free over A. It now follows from Lemma 2.2 of Chapter II that $\mathfrak{C}$ is locally free over $\mathfrak{A}$. $\qquad\square$

Recall that for a locally free $\mathfrak{A}$-module M we write (M) for its class in the locally free classgroup $\operatorname{Cl}(\mathfrak{A})$. We now define a map

$$\psi : \operatorname{PH}(\mathfrak{B}) \longrightarrow \operatorname{Cl}(\mathfrak{A})$$

by setting

$$\psi(\mathfrak{C}) = (\mathfrak{C})(\mathfrak{B})^{-1}.$$

In our arithmetic applications, $\mathfrak{B}$ will in fact be free over $\mathfrak{A}$ (not just locally free) and in this case we have simply $\psi(\mathfrak{C}) = (\mathfrak{C})$. In general, the factor $(\mathfrak{B})^{-1}$ ensures that ψ takes the identity $\mathfrak{B}$ of the group $\operatorname{PH}(\mathfrak{B})$ to the identity $(\mathfrak{A})$ of the group $\operatorname{Cl}(\mathfrak{A})$.

3.2. Theorem. *ψ is a group homomorphism.*

Before proving this, we must develop a little more algebraic machinery. Recall that in Chapter I we gave the Hom-description of $\operatorname{Cl}(\mathfrak{A})$ as a certain quotient of the group $\operatorname{Det}(\mathbb{A}G^{\times}) = \operatorname{Hom}_{\Omega}(R_G, J_F)$, where F is a large enough extension of K. We showed there that the class $(M) \in \operatorname{Cl}(\mathfrak{A})$ of a given locally free $\mathfrak{A}$-module M of rank 1 can be described as follows:- for each prime ideal $\mathfrak{p}$ of $\mathfrak{O}$, including the generic prime $\mathfrak{g} = (0)$, pick an element $m_{\mathfrak{p}}$ of $M_{\mathfrak{p}}$ for which $m_{\mathfrak{p}} \mathfrak{A}_{\mathfrak{p}} = M_{\mathfrak{p}}$. For each $\mathfrak{p} \neq \mathfrak{g}$, let $x_{\mathfrak{p}}$ be the unique element of $K_{\mathfrak{p}} G$ such that $m_{\mathfrak{p}} = m_{\mathfrak{g}} x_{\mathfrak{p}}$. Then (M) is represented by $\mathrm{Det}(\underline{x})$, where $\underline{x} = (x_{\mathfrak{p}})_{\mathfrak{p} \neq \mathfrak{g}} \in AG^{\times}$. We must derive a more convenient description of $\underline{x}$ in the case that M is a principal homogeneous space.

For any principal homogeneous space $\mathfrak{C}$ for $\mathfrak{B}$, write $C = \mathfrak{C}K$ and define the *resolvend map* $r = r_{C} : C \longrightarrow CG$ by

$$r(c) = \sum_{g \in G} c^{g} g^{-1}.$$

For $h \in G$ we have

$$r(c^{h}) = \sum_{g \in G} c^{hg} g^{-1} = \sum_{k \in G} c^{k} k^{-1} h = r(c)\, h,$$

so r is G-equivariant, where G acts on CG by multiplication.

For each prime ideal $\mathfrak{p}$ of $\mathfrak{O}$ pick a basis $c_{\mathfrak{p}}$ for $\mathfrak{C}_{\mathfrak{p}}$ over $\mathfrak{A}_{\mathfrak{p}}$. Then as above $(\mathfrak{C})$ is represented by $\mathrm{Det}(\underline{x})$, with $c_{\mathfrak{p}} = c\, x_{\mathfrak{p}}$ where $c = c_{\mathfrak{g}}$. Taking resolvends,

$$r(c_{\mathfrak{p}}) = r(c\, x_{\mathfrak{p}}) = r(c)\, x_{\mathfrak{p}}$$

and hence

(8)
$$\mathrm{Det}(r(c_{\mathfrak{p}})) = \mathrm{Det}(r(c))\, \mathrm{Det}(x_{\mathfrak{p}}).$$

3.3. Lemma.
$$r(c) \in CG^{\times}.$$

PROOF: We must show that there exist elements y_{h} of C for $h \in G$ such that

$$\left(\sum_{g \in G} c^{g}\, g^{-1} \right) \left(\sum_{h \in G} y_{h}\, h \right) = 1.$$

This amounts to showing that the system of simultaneous equations

(9)
$$\sum_{h \in G} c^{hk^{-1}} y_{h} = \begin{cases} 1 & \text{if } k = 1 \\ 0 & \text{otherwise} \end{cases}$$

is soluble. By definition of $c = c_{\mathfrak{g}}$, $(c^{h})_{h \in G}$ is a K-basis of C, and as C is a separable K-algebra,

$$\det((c^{hk^{-1}})_{h,k})^{2} = \det(\mathrm{Tr}(c^{gh})_{g,h}) \in K^{\times}.$$

Thus $\det((c^{hk^{-1}})_{h,k}) \in C^{\times}$, and (9) is indeed soluble. $\square$

Combining (8) and Lemma 3.3, we have shown

3.4. Lemma. ($\mathfrak{C}$) *is represented by* $\text{Det}(r(c_{\mathfrak{p}})_{\mathfrak{p} \neq \mathfrak{g}}) \, \text{Det}(r(c))^{-1}$.

We also need to evaluate resolvends for products of principal homogeneous spaces. If C and D are two principal homogeneous spaces for B, let

$$\mu : CG \otimes_K DG \longrightarrow (C \otimes_K D)G$$

be the map induced by multiplication in G, and let

$$\text{Tr}_{\delta(G)} : C \otimes D \longrightarrow C.D = (C \otimes D)^{\delta(G)}$$

be the trace map given by

$$\text{Tr}_{\delta(G)}(c \otimes d) = \sum_{g \in G} c^g \otimes d^{g^{-1}}.$$

3.5. Lemma. *For $c \in C$ and $d \in D$,*

$$r_{C.D}(\text{Tr}_{\delta(G)}(c \otimes d)) = \mu(r_C(c) \otimes r_D(d)).$$

PROOF:

$$
\begin{aligned}
r_{C.D}(\text{Tr}_{\delta(G)}(c \otimes d)) &= r_{C.D}\left(\sum_{h \in G} c^h \otimes d^{h^{-1}} \right) \\
&= \left(\sum_{g,h \in G} c^h \otimes d^{h^{-1}} \right)^g g^{-1} \\
&= \sum_{g,h \in G} c^h \otimes d^{h^{-1}g} \, g^{-1} \\
&= \sum_{h,k \in G} c^h \otimes d^k \, k^{-1} \, h^{-1} \\
&= \mu\left(\left(\sum_{h \in G} c^h \, h^{-1} \right) \otimes \left(\sum_{k \in G} d^k \, k^{-1} \right) \right) \\
&= \mu(r_C(c) \otimes r_D(d)).
\end{aligned}
$$

$\square$

PROOF OF THEOREM 3.2: Let $\mathfrak{C}$ and $\mathfrak{D}$ be two principal homogeneous spaces for $\mathfrak{B}$. We must show that $\psi(\mathfrak{C}) \, \psi(\mathfrak{D}) = \psi(\mathfrak{C}.\mathfrak{D})$, i.e. that

(10) $(\mathfrak{C}) \, (\mathfrak{D}) = (\mathfrak{C}.\mathfrak{D}) \, (\mathfrak{B})$

in $\text{Cl}(\mathfrak{A})$.

By Theorem 4.3 of Chapter II, we have the isomorphisms of $\mathfrak{A}$-modules

$$\mathfrak{B} \cong J \otimes \mathfrak{A} \cong \mathfrak{b}\,\mathfrak{A}$$

where $J = \mathfrak{b}\,\ell$ is the ideal of integrals in $\mathfrak{B}$. For each prime ideal $\mathfrak{p}$ of $\mathfrak{O}$, pick a generator $b_\mathfrak{p}$ of the $\mathfrak{O}_\mathfrak{p}$-ideal $\mathfrak{b}\,\mathfrak{O}_\mathfrak{p}$, taking $b_\mathfrak{g} = 1$. Then, writing

$$\underline{b} = (b_\mathfrak{p})_{\mathfrak{p} \neq \mathfrak{g}} \in \mathbb{A}^\times \subseteq \mathbb{A}G^\times,$$

$\mathrm{Det}(\underline{b})$ represents $(\mathfrak{B})$.

Recall from Proposition 4.5 of Chapter II that $\mathfrak{B}$ has inverse different $\mathfrak{b}^{-1}\mathfrak{B}$, so that $\mathrm{Tr}(\mathfrak{B}) = \mathfrak{b}$. Since for a suitable extension L of K we have an isomorphism

$$\mathfrak{C} \otimes \mathfrak{O}_L \cong \mathfrak{B} \otimes \mathfrak{O}_L$$

of $\mathfrak{O}_L$-algebras, it follows that $\mathrm{Tr}(\mathfrak{C}) = \mathfrak{b}$, and similarly $\mathrm{Tr}(\mathfrak{D}) = \mathrm{Tr}(\mathfrak{C}.\mathfrak{D}) = \mathfrak{b}$. Writing $\mathrm{Tr}_{G \times G}$ for the obvious trace map $\mathfrak{C} \otimes \mathfrak{D} \longrightarrow \mathfrak{O}$, we have

$$
\begin{aligned}
\mathrm{Tr}(\mathrm{Tr}_{\delta(G)}(\mathfrak{C} \otimes \mathfrak{D})) &= \mathrm{Tr}_{G \times G}(\mathfrak{C} \otimes \mathfrak{D}) \\
&= \mathrm{Tr}(\mathfrak{C}) \otimes \mathrm{Tr}(\mathfrak{D}) \\
&= \mathfrak{b}^2,
\end{aligned}
$$

and it follows that

(11) $$\mathrm{Tr}_{\delta(G)}(\mathfrak{C} \otimes \mathfrak{D}) = \mathfrak{b}\,\mathfrak{C}.\mathfrak{D}.$$

Now for each $\mathfrak{p}$ choose a generator $c_\mathfrak{p}$ of $\mathfrak{C}_\mathfrak{p}$ over $\mathfrak{A}_\mathfrak{p}$. Set $c = c_\mathfrak{g}$ and $\underline{c} = (c_\mathfrak{p})_{\mathfrak{p} \neq \mathfrak{g}}$. Similarly define d and $\underline{d}$ for $\mathfrak{D}$. Then by Lemma 3.4, $\mathfrak{C}$ and $\mathfrak{D}$ are represented by $\mathrm{Det}(r(\underline{c}))\,\mathrm{Det}(r(c))^{-1}$ and $\mathrm{Det}(r(\underline{d}))\,\mathrm{Det}(r(d))^{-1}$ respectively. On the other hand, it follows from Lemma 3.4, Lemma 3.5 and (11) that $\mathfrak{C}.\mathfrak{D}$ is represented by

$$
\mathrm{Det}(\underline{b}^{-1}\,\mu(r(\underline{c}) \otimes r(\underline{d})))\,\mathrm{Det}(\mu(r(c) \otimes r(d)))^{-1} =
$$
$$
\mathrm{Det}(\underline{b})^{-1}\,\mathrm{Det}(\underline{c})\,\mathrm{Det}(\underline{d})\,\mathrm{Det}(c)^{-1}\,\mathrm{Det}(d)^{-1}.
$$

Thus equation (10) holds, either side being represented by

$$\mathrm{Det}(\underline{c})\,\mathrm{Det}(\underline{d})\,\mathrm{Det}(c)^{-1}\,\mathrm{Det}(d)^{-1}.$$

$\square$

§4   The Kummer Isomorphism

We now take G to be a group of prime order p, and K to be a number field containing a primitive pth root of 1. Let $\mathfrak{A}$ be a Hopf order in $A = KG$, let $\mathfrak{f}$ be its conductor, and as usual let $\mathfrak{B} = \mathfrak{A}^D$. By the Kummer isomorphism of Example 1.4, we may identify $\mathrm{PH}(B)$ with $K^\times / K^{\times\,p}$. We will now describe $\mathrm{PH}(\mathfrak{B})$ and $\ker \psi$ as subgroups of $K^\times / K^{\times\,p}$.

4.1. Definition. *(a)* $\qquad V_{\mathfrak{f}} = \{v \in K^{\times} \;:\; v \in K_{\mathfrak{p}}^{\times\, p}\,(1+\mathfrak{f}^{p}\mathfrak{O}_{\mathfrak{p}}) \text{ for all } \mathfrak{p}\}$

(where we adopt the usual convention that if $\mathfrak{p}$ does not divide $\mathfrak{f}$ then $1+\mathfrak{f}^{p}\mathfrak{O}_{\mathfrak{p}}$ is to be understood as $\mathfrak{O}_{\mathfrak{p}}^{\times}$).

(b) $\qquad Y_{\mathfrak{f}} = \{u \in \mathfrak{O}^{\times} \;:\; u \equiv 1 \pmod{\mathfrak{f}^{p}\mathfrak{O}}\}.$

4.2. Theorem. *Under the Kummer isomorphism $\theta : \mathrm{PH}(B) \longrightarrow K^{\times}/K^{\times\, p}$, we have*

$$\mathrm{PH}(\mathfrak{B}) = \frac{V_{\mathfrak{f}}}{K^{\times\, p}} \text{ and } \ker \psi = \frac{Y_{\mathfrak{f}}\,K^{\times\, p}}{K^{\times\, p}}.$$

4.3. Remark. This reduces the problem of determining $\ker \psi$ to the standard (but difficult) question of finding units satisfying certain congruence conditions.

PROOF OF THEOREM 4.2 (SKETCH): For simplicity we assume that $\mathfrak{B}$ is free over $\mathfrak{A}$, so $\mathfrak{B} = b\,\ell.\mathfrak{A}$ for some $b \in \mathfrak{O}$.

We give in full the proof that

$$\ker \psi \subseteq \frac{Y_{\mathfrak{f}}\,K^{\times\, p}}{K^{\times\, p}}.$$

Let $\mathfrak{C} = c.\mathfrak{A} \in \ker \psi$ and set $C = \mathfrak{C}K$. We may suppose that $\mathfrak{C}$ is not isomorphic to $\mathfrak{B}$ as a principal homogeneous space. Then $C \neq B$ by Lemma 2.1, so C is a Galois field extension of K of degree p. We have to show that $C = K(u)$ for some u with $u^{p} \in Y_{\mathfrak{f}}$.

By Lemma 1.3, we may take as our splitting isomorphism the map

$$\xi : C \otimes C \longrightarrow B \otimes C = \mathrm{Map}(G, C)$$

given by
$$(12) \qquad\qquad \xi(c \otimes d)(g) = c^{g^{-1}} d.$$

Thus, at the integral level, we have an isomorphism

$$\xi : \mathfrak{C} \otimes \mathfrak{O}_{C} \longrightarrow \mathfrak{B} \otimes \mathfrak{O}_{C}$$

of $\mathfrak{A} \otimes \mathfrak{O}_{C}$-modules. It follows from Theorem 4.3 of Chapter II that $\xi(c) = b\,\ell.a$ for some $a \in (\mathfrak{A} \otimes \mathfrak{O}_{C})^{\times}$. Forming resolvents, we have $r(\xi(c)) = r(b\,\ell.a) = b\,r(\ell).a$. Expanding this gives

$$(13) \qquad \sum_{g \in G} \xi(c^{g})\, g^{-1} = b \sum_{g \in G} \ell^{g}\, g^{-1} a \in (B \otimes C)G = \mathrm{Map}(G, CG).$$

We evaluate this map at $1 \in G$:- firstly, it follows from (12) that

$$\xi(c^{g})(1) = \xi(c^{g} \otimes 1)(1) = c^{g},$$

so the left-hand side of (13) yields

$$\sum_{g\in G} c^g\, g^{-1};$$

and secondly we have

$$(\ell^g\, g^{-1}\, a)(1) = \ell^g(1)\, g^{-1}\, a = \begin{cases} a & \text{if } g = 1 \\ 0 & \text{otherwise,} \end{cases}$$

so the right-hand side of (13) gives $b\,a$. We have therefore shown that

$$a = \frac{1}{b}\sum_{g\in G} c^g\, g^{-1}.$$

Now take χ to be a nontrivial character of G. As $a \in (\mathfrak{A}\otimes\mathfrak{O}_C)^\times$, we have $\chi(a),\ \epsilon(a) \in \mathfrak{O}_C{}^\times$. We will show that the unit $u = \chi(a)\,\epsilon(a)^{-1}$ is a generator of the extension C of K and has the required properties.

Firstly, for $h \in G$ we have

$$\begin{aligned}
\chi(a)^h &= \left(\frac{1}{b}\sum_{g\in G} c^g\,\chi(g^{-1})\right)^h \\
&= \frac{1}{b}\sum_{g\in G} c^{gh}\,\chi(g^{-1}) \\
&= \chi(a)\,\chi(h)
\end{aligned}$$

and similarly, $\epsilon(a)^h = \epsilon(a)$. Thus $u^p \in \mathfrak{O}^\times$ but $u \notin K$. Also, $\epsilon(a) \in \mathfrak{O}^\times$.

Secondly,

$$u = \frac{\chi(a)}{\epsilon(a)} = 1 + \frac{\chi(a) - \epsilon(a)}{\epsilon(a)} \in 1 + \mathfrak{f}\mathfrak{O}_C,$$

and since $\mathfrak{f}^{p-1}$ divides p, it follows from the binomial theorem that

$$u^p \in \mathfrak{O}^\times \cap (1 + \mathfrak{f}^p\mathfrak{O}_C) = Y_{\mathfrak{f}}.$$

This completes the proof that

$$\ker\psi \subseteq \frac{Y_{\mathfrak{f}}\,K^{\times p}}{K^{\times p}}.$$

Now let $\mathfrak{C} \in \mathrm{PH}(\mathfrak{B})$. For each prime $\mathfrak{p} \neq (0)$, take an element $c_{\mathfrak{p}} \in \mathfrak{C}_{\mathfrak{p}}$ for which $\mathfrak{C}_{\mathfrak{p}} = c_{\mathfrak{p}}.\mathfrak{A}_{\mathfrak{p}}$. Then we can apply the preceding argument in this local situation to obtain a $u_{\mathfrak{p}}$ satisfying

$$u_{\mathfrak{p}}{}^p \in \mathfrak{O}_{\mathfrak{p}}{}^\times \cap (1 + \mathfrak{f}^p\mathfrak{O}_{C,\mathfrak{p}}) = 1 + \mathfrak{f}^p\mathfrak{O}_{\mathfrak{p}}.$$

Taking a normal basis c for C over KG, we similarly obtain a $v^p \in K^\times$ with $C = K(v)$, and comparing each $u_\mathfrak{p}$ with v, we find that

$$PH(\mathfrak{B}) \subseteq \frac{V_\mathfrak{f}}{K^{\times p}}.$$

Conversely, given $v \in V_\mathfrak{f}$, for each $\mathfrak{p}$ we have $v = v_\mathfrak{p}{}^p\, w_\mathfrak{p}$ with $v_\mathfrak{p} \in K_\mathfrak{p}{}^\times$ and $w_\mathfrak{p} = 1 + \mathfrak{f}^p\mathfrak{O}_\mathfrak{p}$. Let $y_\mathfrak{p} = w_\mathfrak{p}{}^{1/p}$ and let

$$\mathfrak{C}_\mathfrak{p} = b\,y_\mathfrak{p}\,\ell.\mathfrak{A}_\mathfrak{p} \subseteq K_\mathfrak{p}(y_\mathfrak{p})$$

(with the obvious modification if $w_\mathfrak{p} \in K_\mathfrak{p}^{\times\,p}$). Taking $C = K(v^{1/p})$, one can then show that $\cap_\mathfrak{p}(C \cap \mathfrak{C}_\mathfrak{p})$ is a principal homogeneous space for $\mathfrak{B}$, whose image under the Kummer isomorphism θ is the class of v. Moreover, if $v \in Y_\mathfrak{f}$ then $\mathfrak{C} \in \ker \psi$. For the details, see [28]. $\qquad\square$

IV Arithmetic Applications:- The Cyclotomic Case

From now on, we fix an odd prime p and take K to be a number field containing a primitive pth root ζ of 1, and G to be a cyclic group of order p. We identify G with the group of pth roots of 1 in K. As before, we write $A = KG$, $B = \mathrm{Map}(G, K)$, and take $\mathfrak{A}$ and $\mathfrak{B} = \mathfrak{A}^D$ to be Hopf orders in these Hopf algebras. Our aim is to use the properties of certain L-functions arising in arithmetic, together with Theorem 4.2 of Chapter III, to investigate the kernel of the map $\psi : \mathrm{PH}(\mathfrak{B}) \longrightarrow \mathrm{Cl}(\mathfrak{A})$ for certain special Hopf orders $\mathfrak{A}$. Our results therefore concern those principal homogeneous spaces for $\mathfrak{B}$ which are isomorphic to $\mathfrak{B}$ as $\mathfrak{A}$-modules.

In this chapter, the Hopf order $\mathfrak{A}$ in question is the group ring $\mathfrak{O}G$, and the L-functions are p-adic L-functions associated to Dirichlet characters of conductor p. In the final chapter, the Hopf orders and L-functions both arise from an elliptic curve with complex multiplication.

We fix once and for all an embedding $\iota : \mathbb{Q}^c \hookrightarrow \mathbb{Q}_p{}^c$. Any function f taking values in a number field can then be viewed via ι as a taking p-adic values. In particular, if f takes values in the field $\mathbb{Q}(\zeta_{p-1})$ of $(p-1)$th roots of 1, then we can view f as taking values in $\mathbb{Q}_p$.

§1 p-adic L-Functions

We now recall some standard facts about p-adic L-functions. For proofs and further details, see [30].

Let $\chi : (\mathbb{Z}/p\mathbb{Z})^\times \longrightarrow \mathbb{Q}(\zeta_{p-1})^\times$ be a nontrivial Dirichlet character. We say that χ is *odd* or *even* according as $\chi(-1) = -1$ or $+1$. Recall that the Dirichlet L-function is defined by

$$L(s, \chi) = \prod_{q \neq p}(1 - \chi(q)q^{-s})^{-1}$$

where s is a complex variable, and where the product is over all prime numbers $q \neq p$. This converges for $\mathrm{Re}(s) > 1$, and satisfies a functional equation by means of which $L(s, \chi)$ is defined for all $s \in \mathbb{C}$. Hurwitz evaluated $L(1 - n, \chi)$ for positive integers n, showing that this value always lies in $\mathbb{Q}(\zeta_{p-1})$, and is zero whenever n and χ have different parities.

Using the embedding ι, we view χ as a p-adic character, taking values in $\mathbb{Z}_p$. Then χ is some power of the Teichmüller character ϕ, defined by the condition $\phi(a) \equiv a \pmod{p\mathbb{Z}_p}$. To obtain a p-adic analogue of $L(s, \chi)$, we must intertwine the complex L-functions for all the translates of χ by powers of ϕ:

1.1. Theorem. *(Leopoldt, Kubota:- see [30, Theorem 5.11]) For each nontrivial Dirichlet character χ of conductor p, there exists a unique function $L_p(z, \chi)$,* *continuous in $z \in \mathbb{Z}_p$ and taking values in $\mathbb{Q}_p$, such that for all integers $n > 0$*

$$L_p(1-n,\chi) = \begin{cases} L(1-n,\chi\phi^{-n}) & \text{if } \chi \neq \phi^n \\ (1-p^{n-1})\zeta(1-n) & \text{if } \chi = \phi^n \end{cases}$$

(where $\zeta(s)$ denotes the Riemann zeta function).

Note that if $n \equiv 0 \pmod{p-1}$ then the above formula reduces to

$$L_p(1-n,\chi) = L(1-n,\chi).$$

The value $L_p(1,\chi)$ is not specified in Theorem 1.1, but in fact if χ is even then we have the following p-adic limit formula:

1.2. Theorem. *([30, Theorem 5.18]) Let χ be an even, nontrivial Dirichlet character of conductor p. Then*

$$L_p(1,\chi) = -\frac{\tau(\chi)}{p}\sum_{a=1}^{p-1}\log_p(1-\zeta^a)\,\chi(a)^{-1},$$

where $\zeta = \zeta_p$ is a (fixed) primitive pth root of unity,

$$\tau(\chi) = \sum_{a=1}^{p-1}\chi(a)\,\zeta^a$$

and $\log_p$ is the p-adic logarithm.

(Recall that $\log_p$ is uniquely defined on the whole of $\mathbb{Q}_p^{\times}$ by stipulating that $\log_p(p) = 0$, that $\log_p(ab) = \log_p(a)+\log_p(b)$ for all a, b, and that $\log_p(z)$ is given by the usual power series in its domain of p-adic convergence $z \in 1+p\,\mathbb{Z}_p$.)

Theorem 1.2 is analogous to the classical limit formula for the value of the complex L-function at $s = 1$ ([30, Theorem 4.9]).

§2   The i-parts of $\ker \psi$

We now return to the field $K = \mathbb{Q}(\zeta_p)$. Let $\Delta = \mathrm{Gal}(K/\mathbb{Q})$. This is naturally isomorphic to $\mathbb{F}_p^{\times}$, and composing this isomorphism with the Teichmüller character ϕ, we have a canonical character

$$\kappa : \Delta \cong \mathbb{F}_p^{\times} \longrightarrow \mathbb{Z}_p^{\times}$$

of Δ, uniquely determined by the property $\zeta^{\delta} = \zeta^{\kappa(\delta)}$ for $\delta \in \Delta$.

If M is any $\mathbb{Z}_p\Delta$-module (so in particular if M is an $\mathbb{F}_p\Delta$-module) then M has a canonical direct sum decomposition

$$M = \bigoplus_{i=1}^{p-1} M.e_i$$

where the e_i are the primitive idempotents of $\mathbb{Z}_p\Delta$, given by

$$e_i = \frac{1}{p-1} \sum_{\delta \in \Delta} \kappa^{-i}(\delta)\, \delta.$$

We will write M_i for the module $M.e_i$, and refer to this as the i-part of M. The subscripts i should of course be read mod $p-1$. In particular, we can view the Sylow p-subgroup of the ideal classgroup of $\mathfrak{O}$ as a $\mathbb{Z}_p\Delta$-module, and write h_i for the order of its i-part.

We take as our Hopf order $\mathfrak{A}$ the group ring $\mathfrak{O}G$. Thus the conductor of $\mathfrak{A}$ is the prime ideal $\mathfrak{f} = (1 - \zeta)\mathfrak{O}$. Let $\mathfrak{p}$ denote the maximal ideal of the local ring $\mathfrak{O}_{\mathfrak{f}}$. Abusing notation, we will write $K_{\mathfrak{p}}, \mathfrak{O}_{\mathfrak{p}}, Y_{\mathfrak{p}}$ for $K_{\mathfrak{f}}, \mathfrak{O}_{\mathfrak{f}}, Y_{\mathfrak{f}}$ respectively, and will identify $\mathrm{Gal}(K_{\mathfrak{p}}/\mathbb{Q}_p)$ with $\Delta = \mathrm{Gal}(K/\mathbb{Q})$.

By Theorem 4.2 of Chapter III we have

(14)
$$\ker \psi \cong \frac{Y_{\mathfrak{p}} K^{\times p}}{K^{\times p}}.$$

We view $\mathrm{PH}(B) \cong H^1(\Omega, G)$ as an $\mathbb{F}_p\Delta$-module via the action of Δ on $G = \langle \zeta_p \rangle$. Then $\ker \psi$ becomes an $\mathbb{F}_p\Delta$-module, and one can verify that (14) is an isomorphism of $\mathbb{F}_p\Delta$-modules. The main result of this chapter describes the i-parts of $\ker \psi$:

2.1. Theorem. *If i is odd or if $i \equiv 0 \pmod{p-1}$ then $(\ker \psi)_i = 0$.*

If i is even and $i \not\equiv 0 \pmod{p-1}$ then $(\ker \psi)_i = 0$ if and only if $L_p(1, \phi^i)\, h_i^{-1} \not\equiv 0 \pmod{p}$.

(We shall see in the course of the proof that h_i does in fact always divide $L_p(1, \phi^i)$.)

2.2. Remark. In the light of Example 2.2 of Chapter III, we see that Theorem 2.1 is really a result which describes the Galois module structure of the rings of integers of unramified extensions of K of degree p.

The proof of Theorem 2.1 will occupy the rest of this chapter. We start with a lengthy, but fairly standard, analysis of certain groups of local units. We define

$$
\begin{aligned}
U &= \{x \in \mathfrak{O}_{\mathfrak{f}}^{\times} : x \equiv 1 \pmod{\mathfrak{p}}\} & \text{(the local 1-units)}\\
U^{(p)} &= \{x \in \mathfrak{O}_{\mathfrak{f}}^{\times} : x \equiv 1 \pmod{\mathfrak{p}^p}\} & \text{(the local p-units)}\\
Y &= \mathfrak{O}^{\times} \cap U & \text{(the global 1-units)}\\
Y^{(p)} &= \mathfrak{O}^{\times} \cap U^{(p)} & \text{(the global p-units)}.
\end{aligned}
$$

Thus, in the notation of Theorem 4.2 of Chapter III, we have $Y^{(p)} = Y_{\mathfrak{p}}$.

2.3. Lemma. *For $j \geq 0$, $\mathfrak{p}^j/\mathfrak{p}^{j+1} \cong \mathbb{F}_p e_j$ as $\mathbb{F}_p\Delta$-modules.*

PROOF: $\mathfrak{p}^j/\mathfrak{p}^{j+1}$ is cyclic of order p, generated by $(1-\zeta)^j \pmod{\mathfrak{p}^{j+1}}$. For $\delta \in \Delta$, let $a = \kappa(\delta) \in \mathbb{Z}_p$. Then

$$
\begin{aligned}
(1-\zeta)^{j\,\delta} &= (1-\zeta^a)^j \\
&= (1+\zeta+\cdots+\zeta^{a-1})^j\,(1-\zeta)^j \\
&\equiv a^j(1-\zeta)^j \pmod{\mathfrak{p}^{j+1}} \\
&= \kappa(\delta)^j(1-\zeta)^j.
\end{aligned}
$$

Thus Δ acts on $\mathfrak{p}^j/\mathfrak{p}^{j+1}$ via the character κ^j. $\square$

2.4. Lemma. *For $1 \le i \le p-2$,*

$$
(\mathfrak{D}_{\mathfrak{p}})_i = (\mathfrak{p})_i = \tau(\phi^{-i})\mathbb{Z}_p.
$$

PROOF: By Lemma 2.3, Δ acts on $\mathfrak{D}_{\mathfrak{p}}/\mathfrak{p}$ via the character $\epsilon = \kappa^{p-1}$, so $(\mathfrak{D}_{\mathfrak{p}})_i = (\mathfrak{p})_i$ for $1 \le i \le p-2$. Now

$$
\mathfrak{D}_{\mathfrak{p}} = \zeta\,\mathbb{Z}_p\Delta = (p-1)\zeta\,\mathbb{Z}_p\Delta,
$$

so

$$
\begin{aligned}
\mathfrak{D}_{\mathfrak{p}}\,e_i &= \zeta\,(p-1)\,e_i\,\mathbb{Z}_p \\
&= \sum_{a=1}^{p-1} \zeta^a\,\phi^{-i}(a)\,\mathbb{Z}_p \\
&= \tau(\phi^{-i})\,\mathbb{Z}_p.
\end{aligned}
$$

$\square$

2.5. Lemma. *If $1 \le i \le p-2$ then $\tau(\phi^{-i}) \notin p\,\mathfrak{D}_{\mathfrak{p}}$ and*

$$
\tau(\phi^{-i})\,\mathbb{Z}_p \cap p\,\mathfrak{p} \subseteq p\,\tau(\phi^{-i})\,\mathbb{Z}_p.
$$

PROOF: The elements ζ^a for $1 \le a \le p-1$ form a $\mathbb{Z}_p$-basis for $\mathfrak{D}_{\mathfrak{p}}$, and the coefficients $\phi^{-i}(a)$ in the expansion of

$$
\tau(\phi^{-i}) = \sum_{a=1}^{p-1} \phi^{-i}(a)\,\zeta^a
$$

in terms of this basis are not divisible by p. This gives the first assertion, and shows that the valuation v of $\tau(\phi^{-i})$ in $K_{\mathfrak{p}}$ satisfies $v < p-1$. (In fact $v = i$ by [30, Proposition 6.13], but we do not need this.) Since the valuation in $K_{\mathfrak{p}}$ of any element of $\tau(\phi^{-i})\,\mathbb{Z}_p$ is congruent to v mod $p-1$, the second assertion follows. $\square$

2.6. Lemma. *For $j \ge 2$, the p-adic logarithm induces an isomorphism of $\mathbb{Z}_p\Delta$-modules*

$$
\log_p : 1 + \mathfrak{p}^j \longrightarrow \mathfrak{p}^j.
$$

PROOF: For $x \in \mathfrak{p}^j$, the usual power series

$$\log(1+x) = \sum_{r=1}^{\infty} (-1)^{r+1} \frac{x^r}{r}, \quad \left(\text{respectively } \exp(x) = \sum_{r=0}^{\infty} \frac{x^r}{r!} \right)$$

converges p-adically to an element of $\mathfrak{p}^j$ (respectively $1 + \mathfrak{p}^j$), so the logarithm gives an isomorphism of abelian groups ($\mathbb{Z}$-modules) from the multiplicative group $1 + \mathfrak{p}^j$ to the additive group $\mathfrak{p}^j$ (cf. [30, Proposition 5.7]). By continuity, this extends to an isomorphism of $\mathbb{Z}_p$-modules, and if $\delta \in \Delta$ then

$$
\begin{aligned}
\log_p (1+x)^\delta &= \left(\sum_{r=1}^{\infty} (-1)^{r+1} \frac{x^r}{r} \right)^\delta \\
&= \sum_{r=1}^{\infty} (-1)^{r+1} \frac{(x^\delta)^r}{r} \\
&= \log_p (1 + x^\delta)
\end{aligned}
$$

since δ acts as a continuous field automorphism on $K_\mathfrak{p}$. Thus the logarithm is indeed an isomorphism of $\mathbb{Z}_p \Delta$-modules. □

2.7. Lemma. *For $2 \leq i \leq p-2$, U_i is $\mathbb{Z}_p$-cyclic; explicitly,*

$$\log_p(U_i) = \tau(\phi^{-i}) \, \mathbb{Z}_p.$$

PROOF: $U_i = (1 + \mathfrak{p})_i = (1 + \mathfrak{p}^2)_i$ for the relevant values of i, by Lemma 2.3. Now by Lemma 2.6, $\log_p : 1 + \mathfrak{p}^2 \longrightarrow \mathfrak{p}^2$ is an isomorphism of $\mathbb{Z}_p \Delta$-modules, and so takes i-parts to i-parts. Thus

$$
\begin{aligned}
U_i &\cong \log_p(U_i) \\
&= (\log_p(1 + \mathfrak{p}^2))_i \\
&= (\mathfrak{p}^2)_i \\
&= \tau(\phi^{-i}) \, \mathbb{Z}_p.
\end{aligned}
$$

□

2.8. Lemma. $U^{(p)}{}_i \supseteq U_i{}^p$ *for all i, with equality for $2 \leq i \leq p-2$.*

PROOF: By the binomial theorem, we have $U^p \subseteq U^{(p)}$, and as $(U_i)^p = (U^p)_i$ for all i, this gives the first assertion. Now let $2 \leq i \leq p-2$. Then

$$
\begin{aligned}
\log_p(U^{(p)}{}_i) &= \log_p(U_i \cap (1 + p\mathfrak{p})) \\
&= \log_p(U_i) \cap \log_p(1 + p\mathfrak{p}),
\end{aligned}
$$

so, using Lemmas 2.7 and 2.5, we have

$$\log_p(U^{(p)}{}_i) = \tau(\phi^{-i}) \, \mathbb{Z}_p \cap p\mathfrak{p} \subseteq p\,\tau(\phi^{-i}) \, \mathbb{Z}_p = \log_p(U_i{}^p),$$

whence $U^{(p)}{}_i \subseteq U_i{}^p$ as required. □

Note that Lemma 2.8 gives a description of the pth powers in certain groups of local units in terms of a congruence condition.

Next, we need a result on the global units:

2.9. Lemma. $Y^{(p)} \cap K^{\times p} = Y^p$.

PROOF: Using the binomial theorem, it is clear that $Y^p \subseteq Y^{(p)} \cap K^{\times p}$. Conversely, if

$$y = x^p \in K^{\times p} \cap Y^{(p)} = K^{\times p} \cap \mathfrak{O}^\times \cap U^{(p)}$$

then $x \in \mathfrak{O}^\times$ and $x^p \in 1 + \mathfrak{p}^p$. The second condition implies that $x \in 1 + \mathfrak{p} = U$, so $x \in \mathfrak{O}^\times \cap U = Y$, whence $y \in Y^p$ as required. $\square$

By (14) and Lemma 2.9 we have

$$(\ker \psi)_i = 0 \quad \Leftrightarrow \quad \left(\frac{Y^{(p)} K^{\times p}}{K^{\times p}} \right)_i = 0$$

$$\Leftrightarrow \quad \left(\frac{Y^{(p)}}{Y^{(p)} \cap K^{\times p}} \right)_i = 0$$

(15) $$\Leftrightarrow \quad \left(\frac{Y^{(p)}}{Y^p} \right)_i = 0.$$

We cannot take i-parts of the global unit groups $Y^{(p)}$ and Y^p directly, since they are only $\mathbb{Z}\Delta$-modules, not $\mathbb{Z}_p\Delta$-modules (although their quotient is a $\mathbb{Z}_p\Delta$-module as it has exponent dividing p). We can however consider $Y \otimes \mathbb{Z}_p$, and a fairly easy calculation yields

2.10. Proposition. *([30, Propositions 8.10, 8.13])*

$$(Y \otimes \mathbb{Z}_p)_i \cong \begin{cases} < \zeta > & \text{if } i \equiv 1 \pmod{p-1} \\ \mathbb{Z}_p & \text{if } i \text{ is even and } i \not\equiv 0 \pmod{p-1} \\ 0 & \text{if } i \equiv 0 \pmod{p-1}, \text{ or if } i \text{ is odd and } i \not\equiv 1. \end{cases}$$

Now

(16) $$\left(\frac{Y^{(p)}}{Y^p} \right)_i \cong \left(\frac{Y^{(p)}}{Y^p} \otimes \mathbb{Z}_p \right)_i \cong \left(\frac{Y^{(p)} \otimes \mathbb{Z}_p}{Y^p \otimes \mathbb{Z}_p} \right)_i \cong \frac{(Y^{(p)} \otimes \mathbb{Z}_p)_i}{(Y^p \otimes \mathbb{Z}_p)_i}.$$

It then follows immediately from (15) and Proposition 2.10 that $(\ker \psi)_i = 0$ if $i \equiv 1 \pmod{p-1}$, if i is odd and $i \not\equiv 1 \pmod{p-1}$, or if $i \equiv 0 \pmod{p-1}$. This proves the first statement of Theorem 2.1.

From now on, let i be even with $i \not\equiv 0 \pmod{p-1}$. To handle this case, we must relate Y and $Y^{(p)}$ to their p-adic closures $\overline{Y}$ and $\overline{Y^{(p)}}$ in U. By the continuity of multiplication, the closure of Y^p is $\overline{Y}^p$. We need the following deep result:

2.11. Theorem. *The $\mathbb{Z}_p$-rank of $\overline{Y}$ coincides with the $\mathbb{Z}$-rank of Y (which is $(p-3)/2$ by Dirichlet's Unit Theorem).*

2.12. Remark. This assertion is Leopoldt's Conjecture for K. For abelian extensions of $\mathbb{Q}$ (so in particular for our $K = \mathbb{Q}(\zeta_p)$) it is a consequence of Brumer's p-adic version of Baker's result on linear forms in logarithms (cf. [30], Theorems 5.25, 5.29]), but for arbitrary number fields, the conjecture is still unproven.

It follows easily from Theorem 2.11 that

$$(Y^{(p)} \otimes \mathbb{Z}_p)_i = (\overline{Y^{(p)}})_i \quad \text{and} \quad (Y^p \otimes \mathbb{Z}_p)_i = (\overline{Y}^p)_i.$$

Thus from (15) and (16) we have

(17) $$(\ker \psi)_i = 0 \Leftrightarrow \overline{Y^{(p)}}_i = \overline{Y}^p{}_i,$$

so $(\ker \psi)_i = 0$ precisely when the description of the pth powers of local units given by Lemma 2.8 also applies to the closures of the global units.

2.13. Lemma.

$$(\ker \psi)_i = 0 \Leftrightarrow \overline{Y}_i = U_i.$$

PROOF: After (17), it is enough to show that

(18) $$\overline{Y}_i = U_i \Leftrightarrow \overline{Y^{(p)}}_i = \overline{Y}^p{}_i.$$

Now
(19) $$\overline{Y^{(p)}}_i = \overline{Y}_i \cap U^{(p)}{}_i.$$
Thus if $\overline{Y}_i = U_i$ then, using Lemma 2.8, we have

$$\overline{Y^{(p)}}_i = U^{(p)}{}_i = U_i{}^p = \overline{Y}^p{}_i.$$

Conversely, if $\overline{Y}_i \neq U_i$ then, since $U_i \cong \mathbb{Z}_p$ (Lemma 2.7), we must have $\overline{Y}_i \subseteq U_i{}^p = U^{(p)}{}_i$, the last equality being Lemma 2.8. Thus, using (19), we have $\overline{Y^{(p)}}_i = \overline{Y}_i \neq \overline{Y}^p{}_i$. □

We now have to investigate the index $(U_i : \overline{Y}_i)$. To do this we introduce the group C of cyclotomic units of K. This is generated as a $\mathbb{Z}\Delta$-module by the two units

$$\zeta \quad \text{and} \quad \eta = \frac{1 - \zeta^g}{1 - \zeta},$$

where g is a primitive root mod p. It is known that C has finite index in the full group of units $\mathfrak{O}^\times$.

We form the $\mathbb{Z}_p\Delta$-module $C' = \overline{C \cap U}$. This decomposes into i-parts, and we have the equality

$$(U_i : \overline{Y}_i) = (U_i : C_i')(\overline{Y}_i : C_i')^{-1},$$

all three terms being finite. The last term is known by the celebrated result of Mazur and Wiles ([13]):

2.14. Theorem.

$$(\overline{Y}_i : C_i') = h_i.$$

Thus to complete the proof of Theorem 2.1 it only remains to show that

$$(U_i : C_i') \sim L_p(1, \phi^i),$$

where the notation $a \sim b$ means that $a = ub$ for some p-adic unit u.

By Lemma 2.7 we have the isomorphism

$$\log_p : U_i \longrightarrow \tau(\phi^{-i}) \, \mathbb{Z}_p,$$

and since

$$C_i' = \eta^{e_i \mathbb{Z}_p}$$

it follows that

$$\log_p(C_i') = \log_p(\eta) \, e_i \mathbb{Z}_p.$$

Hence

$$
\begin{aligned}
(U_i : C_i') \, \mathbb{Z}_p &= \tau(\phi^{-i})^{-1} \log_p(\eta) \, e_i \, \mathbb{Z}_p \\
&= \frac{\tau(\phi^i)}{p} \log_p(\eta) \, e_i \, \mathbb{Z}_p
\end{aligned}
$$

since $\tau(\phi^i) \, \tau(\phi^{-i}) = \phi(-1)^i \, p = p$ ([30, Lemma 6.1]).

We calculate

$$
\begin{aligned}
(p-1) \log_p(\eta) e_i &= \sum_{\delta \in \Delta} (\log_p(1 - \zeta^g) - \log_p(1 - \zeta))^\delta \kappa^{-i}(\delta) \\
&= \sum_{a=1}^{p-1} \log_p(1 - \zeta^{ga}) \, \phi^{-i}(a) - \sum_{a=1}^{p-1} \log_p(1 - \zeta^a) \, \phi^{-i}(a) \\
&= (\phi^i(g) - 1) \sum_{a=1}^{p-1} \log_p(1 - \zeta^a) \, \phi^{-i}(a) \\
&\sim \sum_{a=1}^{p-1} \log_p(1 - \zeta^a) \, \phi^{-i}(a).
\end{aligned}
$$

Thus

$$
\begin{aligned}
(U_i : C_i') &\sim \frac{\tau(\phi^i)}{p} \left(\sum_{a=1}^{p-1} \log_p(1 - \zeta^a) \, \phi^{-i}(a) \right) \\
&= -L_p(1, \phi^i)
\end{aligned}
$$

by Theorem 1.2. This completes the proof of Theorem 2.1.

V   Arithmetic Applications:- The Elliptic Case

In this final chapter, we will discuss two ways in which our algebraic machinery gives connections between certain L-functions and the Galois module structure of extensions arising from elliptic curves with complex multiplication. The first of these, which we will describe only briefly, is analogous to the cyclotomic result of the previous chapter, and the second concerns the implications of the conjecture of Birch and Swinnerton-Dyer for Galois module structure. Much, but not all, of the content of this chapter can be extended to abelian varieties with complex multiplication.

§1   Elliptic Curves

We begin by recalling some standard facts about elliptic curves. For further details and proofs, see [21] or [9].

Let K be a number field. An *elliptic curve* E over K is a projective curve of genus 1, defined over K, together with a distinguished K-rational point O on the curve. One can choose coordinate functions x and y so that E is given by the generalised Weierstrass model

$$(20) \qquad y^2 + a_1 xy + a_3 y = x^3 + a_2 x^2 + a_4 x + a_6 \quad (a_i \in K),$$

where the given point O is the unique point at infinity on this affine curve. The point O is then also a point of inflexion. The set $E(K)$ of K-rational points of E has the structure of an abelian group, with O as the identity element. In terms of the model (20), the group operation $+_E$ is uniquely determined by the rule that the three points of intersection of any line with E have sum O. The points of intersection must be counted according to multiplicity, so the tangent to E at a point P meets E three times at P if P is a point of inflexion, and twice otherwise. The Mordell-Weil theorem asserts that the group $E(K)$ is finitely generated, so in particular the quotient of $E(K)$ by its torsion subgroup $E(K)_{\text{tors}}$ is a finitely generated free abelian group. The *rank* of E over K is by definition the rank of this group.

An *endomorphism* of E is a morphism of varieties from the curve E to itself which is also a group homomorphism. The endomorphisms form a ring $\text{End}(E)$ which clearly contains $\mathbb{Z}$, the positive integer m being identified with the map

$$P \mapsto \underbrace{P +_E \cdots +_E P}_{m}.$$

In fact, $\text{End}(E)$ either coincides with $\mathbb{Z}$ or is isomorphic to some order R in some imaginary quadratic number field F. In the latter case we say that E has *complex multiplication* by R. In our applications, we shall be interested in the case where E has complex multiplication by the full ring of integers of F, and where moreover E is defined over F.

If $\mathfrak{p}$ is a prime ideal of the ring of integers $\mathfrak{O}$ of K, we say that E has *good reduction* at $\mathfrak{p}$ if there is a generalised Weierstrass model (20) for E in which the a_i are $\mathfrak{p}$-integral, and such that the curve over the field $\mathfrak{O}/\mathfrak{p}$ given by reducing (20) mod $\mathfrak{p}$ is non-singular (i.e. still has genus 1). E has *everywhere good reduction* if it has good reduction at every prime $\mathfrak{p}$, possibly with different models (20) at different primes.

To obtain our Hopf orders $\mathfrak{A}$ and $\mathfrak{B}$ from an elliptic curve E over K with everywhere good reduction, we will use the existence of a Néron minimal model for E (cf. [9, 14 §2], [1]):

1.1. Definition. *A Néron minimal model for E over K is a smooth group scheme $\mathcal{E}$ over $\mathrm{Spec}(\mathfrak{O})$ such that*

(i) *the generic fibre $\mathcal{E}_K$ of $\mathcal{E}$ coincides with E,*

(ii) *for every smooth group scheme $\mathcal{X}$ over $\mathrm{Spec}(\mathfrak{O})$ there is a functorial bijection $\mathrm{Hom}(\mathcal{X}, \mathcal{E}) \longrightarrow \mathrm{Hom}(\mathcal{X}_K, E)$.*

The key point in this definition is that every map $\phi : \mathcal{X}_K \longrightarrow E$ extends to a map $\phi : \mathcal{X} \longrightarrow \mathcal{E}$. Taking $\mathcal{X} = \mathrm{Spec}(\mathfrak{O})$, we have $E(K) = \mathcal{E}(\mathfrak{O})$, and taking $\mathcal{X} = \mathcal{E}$, any endomorphism of E extends to an endomorphism of $\mathcal{E}$. It is easily seen that $\mathcal{E}$ is unique up to isomorphism.

Associated to any elliptic curve E over K is the Hasse-Weil L-function $L(E/K, s)$ ([9, 16§3]). This is a complex-valued function of the complex variable s, defined as an Euler product convergent for $Re(s) > 3/2$. It is conjectured that $L(E/K, s)$ has an analytic continuation to the whole complex plane. If E has complex multiplication then $L(E/K, s)$ can be interpreted as the product of L-functions of certain Grössencharaktere (continuous idele-class characters) associated to E, and in this case the existence of an analytic continuation of $L(E/K, s)$ follows from the corresponding result for L-functions of Grössencharaktere. One can also form p-adic L-functions for Grössencharaktere, analogous to the p-adic L-functions for cyclotomic characters considered in the previous chapter.

For any elliptic curve E over K, one of the conjectures of Birch and Swinnerton-Dyer asserts that the order of the zero of $L(E/K, s)$ at $s = 1$ equals the rank of $E(K)$, so that, in particular, $L(E/K, s) \neq 0$ if and only if $E(K)$ is finite. Thanks to work of Coates-Wiles, Kolyvagin and Rubin, this is known to be true for many curves with complex multiplication.

§2   The Hopf Orders $\mathfrak{A}$ and $\mathfrak{B}$

From now on, we take E to be an elliptic curve defined over a quadratic imaginary number field F, and with complex multiplication by the full ring of integers $\mathfrak{O}_F$ of F. This implies that F has classnumber 1. Let p be a rational prime which splits in F, say $p = \pi\pi^*$, and such that E has good reduction at $\pi\,\mathfrak{O}_F$.

Writing $\mathbb{Q}^c$ for the algebraic closure of $\mathbb{Q}$ in $\mathbb{C}$, we set

$$E_\pi = \ker(\pi : E(\mathbb{Q}^c) \longrightarrow E(\mathbb{Q}^c)).$$

Then, by a standard result, E_π is a group of order p. We write $G = E_\pi$; the group G will play the role of our Galois group, just as the group of pth roots of 1 did in the cyclotomic case.

Now let $K = F(E_\pi)$ be the field obtained by adjoining to F the x- and y-coordinates of the points in E_π. (This is in fact independent of the choice of the model (20) for E over F.) Any field automorphism of K fixing F clearly gives an automorphism of the group E_π. In fact we have

2.1. Proposition. *(Coates, Wiles [4]) K/F is Galois, E has everywhere good reduction over K, and the natural homomorphism*

$$\kappa : \mathrm{Gal}(K/F) \longrightarrow \mathrm{Aut}(E_\pi) \cong (\mathbb{Z}/p\mathbb{Z})^\times$$

is an isomorphism of groups.

Now let $\mathcal{E}$ be the Néron minimal model of E over K. The endomorphism π of E extends to an endomorphism π of $\mathcal{E}$, and we set $\mathcal{E}_\pi = \ker(\pi)$.

2.2. Proposition. *$\mathcal{E}_\pi$ is a finite flat group scheme over $\mathrm{Spec}(\mathfrak{O})$, and its generic fibre $\mathcal{E}_{\pi,K}$ identifies with $\mathrm{Spec}(B)$, where $B = \mathrm{Map}(G, K)$.*

PROOF: $\mathcal{E}$ is an abelian scheme over $\mathrm{Spec}(\mathfrak{O})$ since E has everywhere good reduction, so by [16, Proposition 20.7], $\mathcal{E}_\pi$ is a finite flat group scheme. Also

$$\mathcal{E}_{\pi,K} = (\mathcal{E}_K)_\pi = E_\pi = G = \mathrm{Spec}(B).$$

$\square$

It follows from Proposition 2.2 that $\mathcal{E}_\pi = \mathrm{Spec}(\mathfrak{B})$ for some Hopf order $\mathfrak{B}$ in B. As usual, we write $\mathfrak{A} = \mathfrak{B}^D$, so $\mathfrak{A}$ is a Hopf order in $A = KG$.

We must determine $\mathfrak{A}$ and $\mathfrak{B}$ explicitly. The inverse different of $\mathfrak{B}$ is known from the work of Tate on p-divisible groups:

2.3. Proposition. *[cf. [23, Proposition 2]]*

$$\mathcal{D}(\mathfrak{B})^{-1} = \frac{1}{\pi}\,\mathfrak{B}.$$

We then have

2.4. Proposition.

$$\mathcal{D}(\mathfrak{A})^{-1} = \frac{1}{\pi^*}\,\mathfrak{A}.$$

PROOF: Recall from Proposition 4.5 of Chapter II that $\mathcal{D}(\mathfrak{B})^{-1} = \mathfrak{b}^{-1}\mathfrak{B}$, where $\mathfrak{b}\,\ell$ is the ideal of integrals of $\mathfrak{B}$. Combining this with Proposition 2.3, we deduce that $\mathfrak{b} = \pi\,\mathfrak{O}_K$. Dually, $\mathcal{D}(\mathfrak{A})^{-1} = \mathfrak{a}^{-1}\mathfrak{A}$, where $\mathfrak{a}$ is as in Proposition 4.4 of Chapter II, and by that same Proposition, $\mathfrak{a}\,\mathfrak{b} = p\,\mathfrak{O}_K$. Since $p = \pi\,\pi^*$, this implies that $\mathfrak{a} = \pi^*\,\mathfrak{O}_K$, giving the result.   □

To identify $\mathfrak{A}$, and hence $\mathfrak{B}$, it is enough to determine each local completion $\mathfrak{A}_\mathfrak{q}$ of $\mathfrak{A}$.

2.5. Proposition.

$$\mathfrak{A}_\mathfrak{q} = \begin{cases} \mathfrak{O}_\mathfrak{q}G & \text{if } \mathfrak{q}|\pi^*\mathfrak{O}_K \\ \mathfrak{M}_\mathfrak{q} & \text{otherwise,} \end{cases}$$

where $\mathfrak{M}_\mathfrak{q}$ denotes the maximal order in $K_\mathfrak{q}G$.

PROOF: If $\mathfrak{q} \nmid \pi^*\mathfrak{O}_K$ then

$$\mathcal{D}(\mathfrak{A}_\mathfrak{q})^{-1} = \frac{1}{\pi^*}\,\mathfrak{A}_\mathfrak{q} = \mathfrak{A}_\mathfrak{q},$$

so $\mathfrak{A}_\mathfrak{q}$ is maximal and separable.

If $\mathfrak{q} \mid \pi^*\mathfrak{O}_K$ then

$$\mathcal{D}(\mathfrak{A}_\mathfrak{q})^{-1} = \frac{1}{\pi^*}\,\mathfrak{A}_\mathfrak{q} = \frac{1}{p}\,\mathfrak{A}_\mathfrak{q},$$

so $\mathcal{D}(\mathfrak{A}_\mathfrak{q}) = p\,\mathfrak{A}_\mathfrak{q}$. But also $\mathcal{D}(\mathfrak{O}_\mathfrak{q}G) = p\,\mathfrak{O}_\mathfrak{q}G$, so the Hopf orders $\mathfrak{A}_\mathfrak{q}$ and $\mathfrak{O}_\mathfrak{q}G$ have the same different. Since $\mathfrak{O}_\mathfrak{q}G$ is the unique minimal Hopf order in $K_\mathfrak{q}G$, it follows that $\mathfrak{A}_\mathfrak{q} = \mathfrak{O}_\mathfrak{q}G$.   □

Now that the Hopf orders $\mathfrak{A}$ and $\mathfrak{B}$ are known, we have as before a class invariant map

$$\psi : \text{PH}(\mathfrak{B}) \longrightarrow \text{Cl}(\mathfrak{A}).$$

More generally, for any finite extension N of K, we can likewise construct a class invariant map

(21) $$\psi_N : \text{PH}(\mathfrak{B}_N) \longrightarrow \text{Cl}(\mathfrak{A}_N),$$

where $\mathfrak{A}_N = \mathfrak{A}\otimes_\mathfrak{O}\mathfrak{O}_N$ and $\mathfrak{B}_N = \mathfrak{B}\otimes_\mathfrak{O}\mathfrak{O}_N$. Again, we can consider the i-parts of the kernel of this map.

§3   The i-parts of $\ker\psi_N$

In this section, we briefly describe the elliptic analogue of Theorem 2.1 of Chapter IV. For this we must impose some extra hypotheses on the prime p.

Since E has complex multiplication, there is a Grössencharakter associated to E/F. Let $\mathfrak{f}$ be its conductor. We take as our split prime $p = \pi\pi^*$ a prime $p > 3$ which is coprime to $\mathfrak{f}$ and which satisfies the condition $\pi + \pi^* \neq 1$.

Let $K = F(E_\pi)$ and $K_* = F(E_{\pi^*})$. Then K and K_* are linearly independent over F, and we take as N their compositum $K K_*$. By the Weil pairing, $N = K(\zeta)$ where ζ is a primitive pth root of 1. Writing $\Delta = \mathrm{Gal}(K_*/F)$, we have an isomorphism $\kappa : \Delta \longrightarrow (\mathbb{Z}/p\mathbb{Z})^\times$ as in Proposition 2.1, and we may identify Δ with $\mathrm{Gal}(N/K)$. This enables us to view $\mathrm{PH}(\mathfrak{B}_N)$ and $\ker \psi_N$ as $\mathbb{F}_p\Delta$-modules, and thus to decompose them into i-parts. The arguments of Chapter IV can be carried through in this case, using a suitable group of elliptic units in place of the cyclotomic units, and replacing the theorem of Mazur-Wiles by a result of Kolyvagin. We have to identify G with the group of pth roots of 1, which twists the action of Δ by κ, so that the i-part in the cyclotomic case corresponds to the $(i-1)$-part in the elliptic case. The final outcome is:

3.1. Theorem. *([29, Theorem 3])* If $i = p - 1$ then $(\ker \psi_N)_{i-1} = 0$. If $i \not\equiv 0 \pmod{p-1}$ then

$$(\ker \psi_N)_{i-1} \begin{cases} = 0 & \text{if } L_{\mathfrak{p},\mathfrak{f}}(\xi^{-i})\, h_i^{-1} \not\equiv 0 \pmod{p} \\ \text{has order } p & \text{otherwise.} \end{cases}$$

Here, $L_{\mathfrak{p},\mathfrak{f}}$ is the p-adic L-function defined on the Grössencharaktere of the Galois group over K of the union of the ray classfields of K with conductor $\mathfrak{f}$ times a power of $\mathfrak{p}$; ξ is the Grössencharakter induced by the action of $\mathrm{Gal}(K/F)$ on E_π; and h_i is the order of the i-part of the Sylow p-subgroup of the ideal classgroup of K.

§4   Torsion Points and the Conjecture of Birch and Swinnerton-Dyer

Multiplication by π on E gives rise to the Kummer exact sequence

$$0 \longrightarrow E_\pi \longrightarrow E(\mathbb{Q}^c) \xrightarrow{\pi} E(\mathbb{Q}^c) \longrightarrow 0,$$

and taking fixed points of $\Omega(K) = \mathrm{Gal}(\mathbb{Q}^c/K)$ yields the exact sequence of Galois cohomology

$$0 \longrightarrow E_\pi \longrightarrow E(K) \xrightarrow{\pi} E(K) \longrightarrow H^1(\Omega(K), E_\pi).$$

Now $H^1(\Omega(K), E_\pi) = \mathrm{Hom}(\Omega(K), G)$ which we may identify with $\mathrm{PH}(B)$. Thus we have an injection

$$\frac{E(K)}{\pi E(K)} \longrightarrow \mathrm{PH}(B).$$

Explicitly, given $P \in E(K)$ with $P \notin \pi E(K)$, let $P' \in E(\mathbb{Q}^c)$ be any point with $\pi P' = P$. Then the field $K(P')$ obtained by adjoining the coordinates of P' to K, is a principal homogeneous space for B in the class corresponding to $P \bmod \pi E(K)$.

To obtain a similar cohomological interpretation of $\mathrm{PH}(\mathfrak{B})$ in terms of E, we must take into account the whole of $\mathrm{Spec}(\mathfrak{O})$, not just its generic point, and so we use flat cohomology over $\mathrm{Spec}(\mathfrak{O})$ ([15, III §4]) instead of Galois cohomology. Flat cohomology over $\mathrm{Spec}(K)$ reduces to the usual Galois cohomology. The endomorphism π of the Néron minimal model $\mathcal{E}$ of E over K gives rise to a Kummer exact sequence

$$0 \longrightarrow \mathcal{E}_\pi \longrightarrow \mathcal{E} \xrightarrow{\pi} \mathcal{E} \longrightarrow 0.$$

Taking global sections over $\mathrm{Spec}(\mathfrak{O})$ yields the exact sequence of flat cohomology

$$0 \longrightarrow \mathcal{E}_\pi \longrightarrow \mathcal{E}(\mathfrak{O}) \xrightarrow{\pi} \mathcal{E}(\mathfrak{O}) \longrightarrow H^1(\mathrm{Spec}(\mathfrak{O}), \mathcal{E}_\pi).$$

Now $H^1(\mathrm{Spec}(\mathfrak{O}), \mathcal{E}_\pi) \cong \mathrm{PH}(\mathfrak{B})$ ([15, p123]). Moreover, the universal property of the Néron minimal model gives an isomorphism $E(K) \longrightarrow \mathcal{E}(\mathfrak{O})$, so we have the composite homomorphism

$$\alpha : E(K) \longrightarrow \frac{E(K)}{\pi E(K)} \cong \frac{\mathcal{E}(\mathfrak{O})}{\pi \mathcal{E}(\mathfrak{O})} \hookrightarrow \mathrm{PH}(\mathfrak{B}).$$

Thus for every K-rational point P on E, we obtain a principal homogeneous space $\alpha(P)$ for $\mathfrak{B}$ (not just for B). Composing with the class invariant map ψ, we have a homomorphism

$$\psi' : E(K) \xrightarrow{\alpha} \mathrm{PH}(\mathfrak{B}) \xrightarrow{\psi} \mathrm{Cl}(\mathfrak{A}).$$

It can often be shown that a given point is in the kernel of ψ' by using elliptic functions to construct an explicit normal basis. In particular, as a special case of [22, Theorem 1], we have:

4.1. Theorem. *If π is coprime to the number of roots of unity in F, then*

$$E(K)_{\mathrm{tors}} \subseteq \ker \psi'.$$

This can be interpreted as a result about the rings of integers of ray classfields of F, and so gives a partial integral version of Kronecker's *Jugendtraum*. It is conjectured that the conclusion of Theorem 4.1 holds without the hypothesis on π, and more generally extends to any abelian variety with complex multiplication by $\mathfrak{O}_F$.

We can of course carry out the construction of the homomorphisms α and ψ' for any finite extension N of K, using the map ψ_N of (21). We then obtain homomorphisms $\alpha_N : E(N) \longrightarrow \mathrm{PH}(\mathfrak{B}_N)$ and $\psi'_N : E(N) \longrightarrow \mathrm{Cl}(\mathfrak{A}_N)$. It can be shown that the diagram

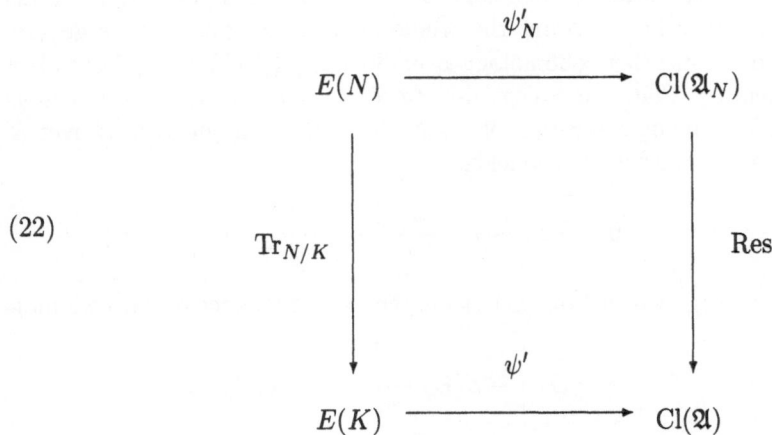

$$(22)$$

commutes, where $\mathrm{Tr}_{N/K}$ is the trace map with respect to the addition on E, and Res is the homomorphism given by restriction of scalars from $\mathfrak{A}_N$ to $\mathfrak{A}$.

Now assume that the conjecture of Birch and Swinnerton-Dyer is known to hold for the curve E over K. If $L(E/K, 1) \neq 0$, then for any finite extension N of K we have the following remarkable chain of implications:

$$
\begin{aligned}
L(E/K, 1) \neq 0 \quad &\Rightarrow \quad E(K) = E(K)_{\mathrm{tors}} && \text{(Birch – Swinnerton-Dyer)} \\
&\Rightarrow \quad \psi' = 0 && \text{(Theorem 4.1)} \\
&\Rightarrow \quad \mathrm{Res} \circ \psi'_N = 0 && \text{(by (22))} \\
&\Rightarrow \quad \text{for any } P \in E(N), \\
&\qquad \alpha_N(P) \cong \mathfrak{A}_N \text{ as } \mathfrak{A}\text{-modules.}
\end{aligned}
$$

Thus, for each algebraic point P (possibly of infinite order) on the curve E, the non-vanishing of $L(E/K, 1)$ gives Galois module stucture information on the ring of integers of the number field obtained from the division of P by π.

References

[1]M.Artin: *Néron Models,* - in G.Cornell, J.H.Silverman (eds.): Arithmetic Geometry. Springer, 1986, pp213-230.

[2]N.P.Byott: *Toroidal Block Decompositions for Hopf Orders in Group Algebras.* to appear in Proc. London Math. Soc.

[3]L.N.Childs, S.Hurley: *Tameness and Local Normal Bases for Objects of Finite Hopf Algebras.* Trans. Am. Math. Soc **298** (1986), 763-778.

[4]J.Coates, A.Wiles: *On the Conjecture of Birch and Swinnerton-Dyer.* Invent. Math. **39** (1977), 223-251.

[5]C.W.Curtis, I.Reiner: Representation Theory of Finite Groups and Associative Algebras. Interscience, 1962, 2nd edn. 1966.

[6]C.W.Curtis, I.Reiner: Methods of Representation Theory, Volume I. John Wiley, 1981, reprinted 1990.

[7]A.Fröhlich: *Invariants for Modules over Commutative Separable Orders.* Quart. J. Math. Oxford (2) **16** (1965), 193-232.

[8]A.Fröhlich: Galois Module Structure of Algebraic Integers. Springer, 1983.

[9]D.Husemöller: Elliptic Curves. (Graduate Texts in Mathematics **111**). Springer 1987.

[10]S.Iitaka: Algebraic Geometry. (Graduate Texts in Mathematics **76**). Springer 1982.

[11]R.G.Larson, M.E.Sweedler: *An associative orthogonal bilinear form for Hopf algebras.* Am. J. Math. **91** (1969), 75-94.

[12]J.Martinet: *Character Theory and Artin L-Functions.* in A.Fröhlich (ed.): Algebraic Number Fields. Academic Press, 1977, pp1-87.

[13]B.Mazur, A.Wiles: *Class fields of abelian extensions of Q.* Invent. Math. **76** (1984), 179-330.

[14]L.R.McCulloh: *Galois module structure of elementary abelian extensions.* J. Algebra **82** (1983), 102-134.

[15]J.S.Milne: Étale Cohomology. Princeton University Press, 1980.

[16]J.S.Milne: *Abelian Varieties* - in G.Cornell, J.H.Silverman (eds.): Arithmetic Geometry. Springer, 1986, pp103-150.

[17]M.Raynaud: *Schémas en groupes de type (p,...,p).* Bull. Soc. Math. France **102** (1974), 241-280.

[18]I.Reiner: Maximal Orders. Academic Press 1975.

[19]J-P.Serre: *Local Class Field Theory* - in J.W.S.Cassels, A.Fröhlich (eds.): Algebraic Number Theory. Academic Press, 1967, pp129-161.

[20]J-P.Serre: Local Fields. (Graduate Texts in Mathematics **67**). Springer 1979.

[21]J.H.Silverman: The Arithmetic of Elliptic Curves. (Graduate Texts in Mathematics **106**). Springer 1986.

[22]A.Srivastav, M.J.Taylor: *Elliptic Curves with Complex Multiplication and Galois Module Structure*. Invent. Math. **99** (1990), 165-184.

[23]J.Tate: *p-divisible Groups* - in T.A.Springer (ed): Proceedings of a Conference on Local Fields (Driebergen, 1966). Springer, 1967, pp158-183.

[24]J.Tate, F.Oort: *Group schemes of prime order*. Ann. Sc. Ec. Norm. Sup. 4^e Serie **3** (1970), 1-21.

[25]M.J.Taylor: *On Fröhlich's Conjecture for Rings of Integers of Tame Extensions*. Invent. Math. **63** (1981), 41-79.

[26]M.J.Taylor: Classgroups of Group Rings. (London Mathematical Society Lecture Notes **91**). Cambridge University Press, 1984.

[27]M.J.Taylor: *Mordell-Weil Groups and the Galois Module Structure of Rings of Integers*. Illinois J. Math **32** (1988), 428-452.

[28]M.J.Taylor: *Résolvendes et Espaces Homogènes Principaux de Schémas en Groupe*. Sém. de Théorie des Nombres, Bordeaux, **2** (1990), 255-271.

[29]M.J.Taylor: *The Galois Module Structure of Certain Arithmetic Principal Homogeneous Spaces* - preprint, 1990.

[30]L.C.Washington: Introduction to Cyclotomic Fields (Graduate Texts in Mathematics **83**). Springer 1982.

[31]W.C.Waterhouse: Introduction to Affine Group Schemes. (Graduate Texts in Mathematics **66**). Springer 1979.